General Relativity: The Essentials

In this short book, renowned theoretical physicist and author Carlo Rovelli gives a straightforward introduction to Einstein's general relativity, our current theory of gravitation. Focusing on conceptual clarity, he derives all the basic results in the simplest way, taking care to explain the physical, philosophical, and mathematical ideas at the heart of 'the most beautiful of all scientific theories'. Some of the main applications of general relativity also are explored, for example, black holes, gravitational waves, and cosmology, and the book concludes with a brief introduction to quantum gravity. Written by an author well known for the clarity of his presentation of scientific ideas, this concise book will appeal to university students looking to improve their understanding of the principal concepts, as well as to science-literate readers who are curious about the real theory of general relativity, at a level beyond a popular science treatment.

CARLO ROVELLI is Director of the Quantum Gravity group at the Centre de Physique Théorique of Aix-Marseille University; he also holds positions at the University of Western Ontario and the Perimeter Institute in Canada. Among his academic contributions in theoretical physics, he is best known as one of the formulators of Loop Quantum Gravity. He has written two monographs for Cambridge University Press, *Quantum Gravity* (2004) and (with Francesca Vidotto) *Covariant Loop Quantum Gravity* (2014). He is also the author of several international bestsellers in popular science such as *Seven Brief Lessons on Physics* (2016) and *The Order of Time* (2017).

General Relativity: The Essentials

CARLO ROVELLI
Université d'Aix-Marseille

CAMBRIDGE
UNIVERSITY PRESS

CAMBRIDGE
UNIVERSITY PRESS

University Printing House, Cambridge CB2 8BS, United Kingdom

One Liberty Plaza, 20th Floor, New York, NY 10006, USA

477 Williamstown Road, Port Melbourne, VIC 3207, Australia

314–321, 3rd Floor, Plot 3, Splendor Forum, Jasola District Centre,
New Delhi – 110025, India

103 Penang Road, #05–06/07, Visioncrest Commercial, Singapore 238467

Cambridge University Press is part of the University of Cambridge.

It furthers the University's mission by disseminating knowledge in the pursuit of
education, learning, and research at the highest international levels of excellence.

www.cambridge.org
Information on this title: www.cambridge.org/9781316516072
DOI: 10.1017/9781009031806

© Adelphi Edizioni S.p.A 2021

First published 2021

A catalogue record for this publication is available from the British Library.

ISBN 978-1-316-51607-2 Hardback
ISBN 978-1-009-01369-7 Paperback

Contents

Preface

There are absolute masterpieces, that move us intensely, Mozart's
Requiem, the Odyssey, the Sistine Chapel, King Lear...To grasp their
splendour may require an apprenticeship. But the prize is pure beauty. And
not only: also the opening of a new look at the world to our eyes.
General relativity, the jewel of Albert Einstein, is one of them.

This short book offers a compact introduction to general relativity,
its conceptual structure, and its basic results.

The focus is on the ideas, without extensive details. The main
results are derived in their simplest form, without lengthy mathematical manipulations. Some original considerations are included,
and some topics are discussed from a perspective not easily found
elsewhere. A final chapter introduces elementary ideas on quantum
gravity. The book can be used to learn key ideas and results of general
relativity without the ambition of becoming fully expert on its vast
ramifications.

It can also be used as a complement to the numerous extensive
manuals,[1] offering additional *conceptual* clarity. It presents general
relativity in the way I understand it today, which I think is the best
perspective to address its quantum aspects.

---

[1] A good and motivated student looks at many books on the same topic. Two classics I still use
as references are Bob Wald's *General Relativity*, which is mathematically oriented, empha-
sises the geometrical perspective, and has a lot of advanced material, and Steven Weinberg's
Gravitation and Cosmology, which de-emphasises geometry. Modern textbooks are Sean
Carroll's *Spacetime and Geometry* and Lewis Ryder's *Introduction to General Relativity*:
these are vastly more comprehensive than the simple introduction given here. A good mathematically oriented text is Yvonne Choquet-Bruhat's *Introduction to General Relativity,
Black Holes, and Cosmology*. In French, a nice introduction is *Relativité générale: Cours et
exercices corrigés* by Aurélien Barrau. This is to mention only the few I know best.

Simple mathematical steps between equations are skipped and indicated with the text *'[do it!]'*. The reader can trust the author, as physicists often do when reading maths; or they can work out the steps and acquire technical competence. A student that does so will end up with good hands-on experience of the technology of relativity. If you like doing exercises, there are several books of exercises in general relativity.[2] The texts in small characters are somewhat marginal with respect to the main topics.

Thanks to Pietropaolo Frisoni for the many corrections and for the translation of the text into Italian. Thanks also to Aymeric Derville for spotting many typos and mistakes in the first draft. I am sure there are more: if you find them, I will be grateful if you can point them out to me.

The recent Nobel Prizes for general relativistic physics (gravitational waves 2017, cosmology 2019, black holes 2020) are a testament to the current vitality and fecundity of this extraordinary theory, Einstein's jewel. Here I try to bring to light the shining beauty and simplicity of the ideas on which it is based.

---

[2] For instance, Thomas A. Moore's *A General Relativity Workbook*.

PART I Bases

WHAT IS GENERAL RELATIVITY?

General relativity is our best current theory describing (i) the gravitational interaction and (ii) the geometrical aspects of space and time. The fact that these two topics go together is a characteristic aspect of the physical content of the theory.

The theory was developed by Albert Einstein, with a little help from a few friends, during roughly 10 years, between 1907 and 1917. Today it finds vast applications in astrophysics and cosmology, and in some technological applications, in particular GPS technology (Global Positioning System), which has changed the way we travel.

The theory has made astonishing predictions. These include black holes, gravitational waves, expansion of the universe, gravitational red shift, and time dilatation. These have *all* been spectacularly confirmed by experiments and observations. So far, it has received only empirical support and has never been found wrong. Many alternative gravitational theories have been studied, but observations over the past century have constantly favoured general relativity, ruling out a large number of alternatives. The most recent occurrence of this was the nearly simultaneous detection in 2017 of gravitational and electromagnetic signals emitted by the merging of two neutron stars; this detection verified with a precision of one part in 10^{15} the general relativity's prediction that the two signals travel at the same velocity, ruling out a large number of other theories that gave different predictions.

The domain of validity of the theory is limited by the fact that it does not account for quantum effects. These are expected at scales of the order of the length

$$L_{Pl} = \sqrt{\frac{\hbar G}{c^3}} \sim 10^{-33}\,\text{cm}, \qquad (0.1)$$

called the Planck length (G is the Newton constant, $\hbar$ the Planck constant, and c the speed of light [*Check that L_{Pl} has indeed dimensions of a length*]). We have only small indirect empirical or observational evidence on this regime, which is likely to be crucial at the centre of black holes, at the end of their evaporation, and in the very early universe. The last chapter of the book mentions tentative ideas on how the theory can be extended to incorporate quantum phenomena.

The theory is based on a simple idea: gravity is described by a field theory like electromagnetism, but this field *also* determines what we call the geometrical properties of spacetime.

The foundations of the theory have three roots: namely in physics, philosophy, and mathematics. The next three chapters examine these roots separately.

I Physics: A Field Theory for Gravity

The first root of general relativity is the spectacular empirical success of Maxwell electromagnetism, which is the basis of our electric and electronic technology.

Maxwell theory is a *field* theory. This means that electric and magnetic interactions are not understood as forces acting at a distance between charges (as in Coulomb), but rather as local interactions carried around at a finite speed by a field: the electromagnetic field.

General relativity does the same for gravity. It does not describe gravity as a force between masses that acts at a distance (as in Newton) but as a local interaction carried around at finite speed by a field: the gravitational field.

General relativity is the field theory of the gravitational field, just as Maxwell theory is the field theory of the electromagnetic field. Maxwell theory has been the principal source of inspiration for Einstein in building general relativity.

The need for a field theory for gravity appeared clear to Einstein because of special relativity. I assume the reader is familiar with the basics of special relativity, and in the next section I spell out in detail why special relativity implies that gravity should be described by a field.

I.I SPECIAL RELATIVITY

- *The physical meaning of Galilean relativity*

Newtonian mechanics is invariant under Galilean transformations such as

$$x' = x - vt, \tag{1.1}$$

which express the fact that position and velocity are relative physical quantities. That is, the position x and the velocity v of an object are only defined with respect to another object (called, in this context, the reference system).

In equation (1.1), x is the 'position', defined as the distance from a reference object O, while x' is the distance from a second reference object O' moving at a constant speed v with respect to O. The quantity t is the time measured by a clock. The invariance of Newtonian mechanics follows from the fact that its fundamental law is

$$F = ma \tag{1.2}$$

and the acceleration $a = d^2x(t)/dt^2$ does not change under (1.1). Indeed the acceleration with respect to O' is $a' = d^2x'(t)/dt^2 = d^2/dt^2$ $(x(t) - vt) = d^2x(t)/dt^2 = a$. Hence if (1.2) is true for the position x defined with respect to O, it also holds true for the position x' defined with respect to O'.

It follows that it is impossible to distinguish uniform rectilinear motion from stasis using mechanical experiments. Position and velocity are only defined as relative to something else.

This implies that it is impossible to label spacetime events with a preferred spatial position variable x.

That is, *given two events happening at different times, it is meaningless to say that they happen 'at the same position x', unless we specify (explicitly or implicitly) a reference object with respect to which position is determined.*

'To remain at the same place' with respect to a moving train, with respect to the Earth, with respect to the Sun, or with respect to the Galaxy, have different meanings. A mother telling 'stop moving' to her child on a train does not mean that the child should jump off the train and stop moving with respect to the Earth. 'To remain at the same place' has no sense, unless we specify with respect to what. See the first two panels of Figure 1.1. This is Galilean relativity.

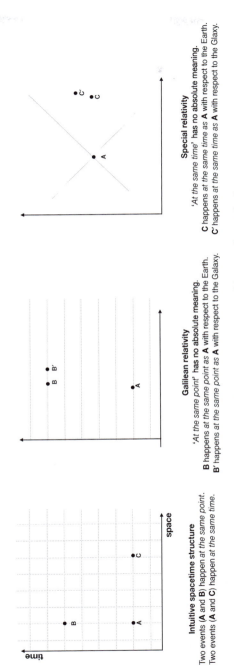

FIGURE 1.1 The structure of spacetime; non-relativistic intuition, Galilean, and special relativity

Intuitive spacetime structure
Two events (**A** and **B**) happen *at the same point*.
Two events (**A** and **C**) happen *at the same time*.

Galilean relativity
'*At the same point*' has **no absolute meaning**.
B happens *at the same point* as **A** with respect to the Earth.
B′ happens *at the same point* as **A** with respect to the Galaxy.

Special relativity
'*At the same time*' has **no absolute meaning**.
C happens *at the same time* as **A** with respect to the Earth.
C′ happens *at the same time* as **A** with respect to the Galaxy.

space

time

- ### *The physical meaning of special relativity*

The Maxwell equations are not invariant under (1.1). Lorentz and Poincaré realised that they are instead invariant under a different set of transformations, such as

$$x' = \gamma(x - vt), \qquad t' = \gamma(t - vx/c^2), \tag{1.3}$$

which we call today Lorentz transformations. Here $\gamma = 1/\sqrt{1 - v^2/c^2}$. While the meaning of x' was clear to Lorentz and Poincaré (it is the distance from a moving object), the meaning of t' remained obscure until Einstein.

In 1905, Einstein clarified this meaning by realising that if t is the time measured by a clock moving together with the reference object O, then an identical clock moving together with the object O' will measure t' rather than t. That is, *identical clocks moving with respect to one another measure different times*. This is what Einstein understood in 1905.

This is not a matter of perspective or definitions. It is a physical fact. Consider two identical clocks separated and then brought back together. Say one of the two moves inertially (no acceleration) between the separation and the reunion and measures the time lapse t between separation and reunion. Say the other clock moves at a (possibly variable) speed v with respect to the first. Then, when they meet again, the second will be in advance of the first. If the square of the speed v of the second clock is constant, the second clock measures the time lapse

$$t' = \frac{1}{\gamma}t < t. \tag{1.4}$$

(If the velocity varies as $v(t)$, then $t' = \int_0^t d\tau \sqrt{1 - v^2(\tau)/c^2} < t$.) The time measured by a clock between two given events depends on the motion of the clock. It is maximal for a clock moving inertially between the two events.

It follows from this property of time intervals that it is impossible to label spacetime events with a unique physically preferred time t. That is, *given two events happening in different locations, it is meaningless to say that two events happen 'at the same time t', unless we specify (explicitly or implicitly) a reference object with respect to which time is determined.*

In other words, 'to happen at the same time' with respect to a moving train, with respect to the Earth, with respect to the Sun, and with respect to the Galaxy, have different meanings. See the third panel of Figure 1.1. To ask what is happening 'now' on Andromeda is a question that has no meaning. This is 'special relativity'.

Notice that it is an extension of Galilean relativity from space to time: Galilean relativity is the discovery that 'to be at the same place' at different times is an ill-defined notion, while special relativity is the discovery that 'to happen at the same time' in different places is an equally ill-defined notion.

1.2   FIELDS

• **Special relativity and Coulomb's law**

A consequence of this discovery is that it makes no sense to say that a force acts at a distance instantaneously: when unqualified, 'instantaneously' is meaningless.

This may seem to be in contradiction to Coulomb's law, which states that two charges e and e' at a distance r act on each other with a repulsive force

$$F = \frac{e\,e'}{r^2}. \tag{1.5}$$

[Before reading ahead, try to answer the following: how can this law be compatible with special relativity?]

If Coulomb's law were a universal law, it would contradict special relativity. But it is not a universal law: it is *only* valid in the static limit in which the charges do not move or move slowly with respect to one another. In this case, they themselves define the reference system where the law holds.

To see why Coulomb's law does not have universal validity, consider what happens if we rapidly take away one of the two charges. Does the other one immediately cease to feel the electric force?

The answer is, of course, negative, because information does not travel faster than light: for a time $t = r/c$, the remaining charge *keeps feeling the force*. During this time, a disturbance of the electromagnetic *field* travels across the space between the charges at the speed of light, and only when it reaches the second charge does the force on it change. Hence, Coulomb's law is compatible with special relativity *only* because it is the static, non-relativistic limit of the interactions carried by a field. This is the observation that motivates general relativity.

• *Special relativity and Newton's law*

Let us come to gravity. According to Newton's law, two masses m and m' at a distance r act on each other with the attractive force

$$F = G \frac{m\, m'}{r^2}.$$

(1.6)

If this was a universal law, it would contradict special relativity. Indeed: what happens if we rapidly move one of the two masses away from the other? Does the other one instantaneously cease to feel the gravitational force? If special relativity is correct, this cannot be, because information does not travel faster than light: for a time $t = r/c$, the remaining mass keeps feeling the force. During this time, a disturbance of a *gravitational* field must travel across space at the speed of light, and only when it reaches the second mass will the gravitational force on it change. For this to be possible, there should exist a field describing the degrees of freedom of what travels between one mass and the other.

Therefore, special relativity implies that Newton's law (1.6) is not a universal law: it is a static, non-relativistic limit, valid only when the masses do not move rapidly with respect to each other. Away from this limit, gravity should be described by a field

	Electromagnetism	Gravity
Static limit	$F = \dfrac{ee'}{r^2}$	$F = G\dfrac{mm'}{r^2}$
	Coulomb's law	Newton's law
Full theory	Maxwell's field theory	General relativity

FIGURE 1.2 The logic that led Einstein to general relativity. Coulomb's law is compatible with special relativity only because it is the static limit of a field theory: Maxwell's electrodynamics. Similarly, to be compatible with special relativity Newton's law must be the static limit of a field theory: general relativity.

theory, capable of accounting for the finite speed of propagation of the interaction. General relativity is such a field theory. See Figure 1.2.

• *The structure of general relativity*

Maxwell theory is defined by (i) a field: the electromagnetic field, (ii) a force law: the equation that describes how charges move in the field, called the Lorentz force equation, and (iii) field equations: the Maxwell equations.

In parallel, as we shall see, general relativity is defined by (i) a field: the gravitational field, (ii) a law that describes how masses move under the action of this field: the 'geodesic equation', and (iii) field equations: the Einstein equations. See Figure 1.3. This is the structure of the theory that I am going to describe in this book.

However, there is an aspect of gravity that makes it sharply different from electromagnetism. The gravitational field is *also* related to the geometrical structure of spacetime. Discovering this connection has been Einstein's everlasting contribution. This is discussed in the next chapter.

	Electromagnetism	General relativity
Field	Maxwell potential $A_a(x)$	Gravitational field $g_{ab}(x)$
Particle eq. of m.	Lorentz force $\ddot{x}^a = \frac{e}{m} F^a{}_b \dot{x}^b$	Geodesic eq. $\ddot{x}^a = -\Gamma^a{}_{bc} \dot{x}^b \dot{x}^c$
Field equations	Maxwell eqs. $D_a F^{ab} = 4\pi J^b$	Einstein eqs. $R_{ab} - \frac{1}{2} R g_{ab} + \lambda g_{ab} = 8\pi G T_{ab}$

FIGURE 1.3 Comparison of the structure of electromagnetism and general relativity. The quantities in the equations will be defined and discussed in the following sections: the gravitational field g_{ab} in Section 3.2; the Levi-Civita connection Γ^a_{bc}, constructed with first derivatives of g_{ab}, in Section 3.2.1; the Ricci tensor R_{ab}, constructed with second derivatives of g_{ab}, in Section 3.2.3; the cosmological constant λ in Section 4.3; and the energy momentum tensor T_{ab} in Chapter 5.

2   Philosophy: What Are Space and Time?

2.1   RELATIVE VERSUS NEWTONIAN SPACE AND TIME

• *The novelty of the Newtonian conception of space and time*

Before Newton, space was understood as the relative arrangement of the things in the world ('The man is here, near the fountain; the deer is there, among the trees.'). Time was generally understood as the counting of the changes in the happenings of the world ('Day, night, day, night,...'). These are *relational* notions of space and time. They are used in common language and they have been the dominant way of understanding these notions in Western philosophy all the way from Aristotle to Descartes.

A consequence of this way of understanding space and time is that there is no space without things, and there is no time if nothing happens, because space is an arrangement of things and time is a counting of happenings.

Newton broke with this tradition. He realised that besides these relational notions of space and time (which he called 'relative'), it is convenient to assume that there is *also* a sort of spatially extended *entity* that exists by itself even if there is nothing else around, and there is a sort of temporally extended *entity* that passes by itself even if nothing else happens. He called 'absolute space' and 'absolute time' these entities that exist by themselves irrespectively from things. We call them 'Newtonian space' and 'Newtonian time'.

• *Structure of Newtonian space*

Newton assumed his space to have the structure of a three-dimensional (3d) Euclidean space, on which we can choose Cartesian coordinates $x^i = (x, y, z)$. He assumed his time to have the metric structure of the real line, coordinated by a variable t. Both the Cartesian coordinates of space and the time variable have *metric* meaning. That is, they correspond to readings of rods and clocks. The length ds of a rod with extremes at x^i and $x^i + dx^i$ is given by

$$ds^2 = dx^2 + dy^2 + dz^2 \equiv \delta_{ij}\, dx^i dx^j. \tag{2.1}$$

Here $\delta_{ij} = \text{diag}[1, 1, 1]$ is the 3×3 identity matrix. In this book the Einstein index convention is understood: each couple of repeated indices, one upstairs and one downstairs, is summed over. Hence $\delta_{ij}\, dx^i dx^j \equiv \sum_{ij} \delta_{ij}\, dx^i dx^j$.

It is important to note that Newton did not deny the relevance of the traditional 'relative' notions of space and time. He simply assumed that *besides* these notions, we should also postulate something else: the existence of these peculiar entities that are Newtonian space and Newtonian time.

Caveat 1: What is 'absolute' in Newtonian mechanics, as in special relativity, is not the position or the velocity of an object; it is its *acceleration*. Newtonian physics requires absolute acceleration to be defined. By 'Newtonian space' or 'Minkowski spacetime' we mean the structures that determine such absolute *acceleration*.

In Newton's original writings, there is some possible ambiguity on the meaning of absolute position and velocity; but the issue has been fully clarified in later developments of Newtonian mechanics. By 'Newtonian space' we do not mean today a preferred reference system; we mean the structure that determines the class of reference systems that are inertial.

Caveat 2: It is sometimes stated that the Newtonian notions of space and time are instinctive and natural. They are not. Perhaps they are now familiar, after centuries of success of Newton's physics. We learn them at school. But they are not natural. Before Newton, the dominant understanding of space and time, both in common use and in the learned tradition, was the relational one.

Euclidean geometry in particular was not interpreted as the geometry of *space*, but rather as the geometry of idealised *objects*.

- **Structure of special relativistic space**

Special relativity is the discovery that Newtonian space and Newtonian time are better described by a single 4d (four-dimensional) geometrical entity: Minkowski space. Its geometry is defined by giving the quantity

$$ds^2 = -dt^2 + dx^2 + dy^2 + dz^2 \equiv \eta_{ab}\, dx^a dx^b \qquad (2.2)$$

which is the square of a distance if positive and minus the square of a proper time if negative. Here $\eta_{ab} = \text{diag}[-1, 1, 1, 1]$ is a 4×4 matrix: the Minkowski metric.

For the discussion that concerns us here, the difference between Newtonian space and time and Minkowski spacetime is not relevant: the two have the same nature. The first is the approximation of the second for slow relative velocities.

• *Nature of Newtonian and special relativistic spacetime*
The nature of both kinds of space has long remained pretty obscure and has given rise to much debate.

Newton characterised space as the 'sensorium of God', whatever that meant. Numerous philosophers, such as Leibniz, Berkeley, and Mach, questioned the cogency of Newton's construction. Kant attempted to understand it as forms a priori necessary for knowledge. Einstein knew these philosophers well and was strongly influenced by their criticisms of the Newtonian conception. By the age of 16, Einstein had already read all three of Immanuel Kant's major works. If you want to do great science, read philosophy.

Newtonian space and time are 'entities' in the sense that they exist irrespectively of anything else but are very different from any other physical entity of the world. For instance, they have no dynamics and cannot be acted upon, even if they determine the way other entities move. They are strange beasts indeed.

What are they actually?

2.2 EINSTEIN'S IDEA: NEWTONIAN SPACE AND TIME ARE A PHYSICAL FIELD

Einstein's great idea at the foundation of general relativity is that Newtonian and special relativistic space and time are indeed real entities – as Newton correctly figured out – but they are not the

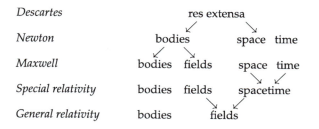

FIGURE 2.1 The evolution of the ontology of physics. This book describes the last step: the identification of physical spacetime with a field. A further step is quantum theory, where bodies and fields are merged as well: all bodies are aspects of (quantum) fields.

strange non-dynamical entities Newton assumed. Rather, they are one of the physical fields that exist in the world: they *are* the gravitational field.

It is the gravitational field that determines the rate at which a clock ticks, or the separation between the ends of a rod (because atoms and moving parts of the clock interact with the gravitational field). Hence what we call the spacetime geometry – read by rods and clocks – is a manifestation of a real and dynamical physical field: the gravitational field.

It follows that Minkowski spacetime, described by the fixed Minkowski metric η_{ab}, is nothing else than the gravitational field in an approximation where we disregard its dynamics.

This remarkable idea changes the set of basic ingredients that we assume form the physical world, reducing Newtonian space and time to a physical field. See Figure 2.1.

Einstein's idea is, therefore, to step out from the approximation where the gravitational field is described by Minkowski space by replacing the *fixed* Minkowski metric η_{ab} with a genuine *field* $g_{ab}(x)$, which varies from point to point. This implies replacing equation (2.2) with the equation

$$ds^2 = g_{ab}(x)\, dx^a dx^b. \tag{2.3}$$

The field $g_{ab}(x)$ takes the constant value $g_{ab}(x) = \eta_{ab}$ only when gravity can be disregarded; in general, it is a genuine *physical field* (with two indices, symmetric) that varies from point to point, governed by field equations and interacting with matter. The field $g_{ab}(x)$ describes the varying geometry of spacetime and at the same time is the gravitational field.

If Minkowski space is only an approximation valid when gravity is negligible, then in general the geometry of spacetime is not Minkowskian. In particular, the geometry of space is not Euclidean. That is, since the geometrical aspects of space and time are determined by the gravitational field, which varies from point to point, they must also be variable, or deformable. In the presence of gravity, the Minkowski geometry of spacetime is deformed.

Einstein was lucky to find the mathematics for describing such 'curved' non-Euclidean and non-Minkowskian spaces already largely developed by the mathematicians, in particular in a mathematical theory developed by Bernhard Riemann. The main equation of Riemannian geometry, which I illustrate in the next chapter, is precisely equation (2.3).

Before concluding this chapter, however, let us see *how* Einstein came to the extraordinary idea that the geometry of spacetime is nothing else than the gravitational field.

Which hints did he have?

2.3   EINSTEIN'S HINT: ACCELERATION WITH RESPECT TO WHAT?

Newton's argument for the existence of absolute space and time is the observation that there are inertial forces. Inertial forces are due to the acceleration of the reference system. Acceleration with respect to what? Newton's answer: with respect to the absolute structure of space.

For Newton (and in special relativity) inertial forces are due to acceleration with respect to the fixed geometry of spacetime.

Newton's bucket, I. In a famous page of the *Principia* (Philosophiæ Naturalis Principia Mathematica (Royal Society 1687), Newton claims that an experiment with a bucket full of water proves the existence of the 'absolute space' he postulates. The surface of the water in

Phase 1:
Water does not rotate in absolute space.
Rotates with respect to bucket.
No concavity of its surface.

Phase 2:
Water rotates in absolute space.
Does not rotate with respect to bucket.
Concavity of its surface.

FIGURE 2.2 Newton's argument for the existence of absolute space: the physical effect (the concavity of the surface of the water) is due to the rotation with respect to absolute space, not to the relative rotation with respect to the container.

the bucket becomes concave if the water rotates around the axis of the bucket. But which rotation does so? Is it rotation with respect to its container that causes the concavity of the water surface? Not so – observes Newton – because if the bucket starts rotating, the water is dragged along by friction to rotate with the bucket *only after some time*. See Figure 2.2. During the first transient period, the bucket spins with respect to us, but the water doesn't yet. During this first transient phase the water *is* in rotation relative to its container, but there is *no* concavity yet. The concavity appears later on, when the water is not anymore in rotation relative to its container. Hence, relative motion with respect to the surroundings (the only true relative motion) has no effect here. What has an effect is the *absolute* rotation of the water: rotation with respect to absolute space. Since it has physical effects – argues Newton – absolute space must be real. The argument is flawless.

Therefore, Newton's space and time are the entities that determine what is accelerating and what is not accelerating.

Einstein, however, is struck by the fact that gravity has a remarkable peculiarity in this respect. Suppose you are in a spaceship orbiting around the Earth. Your motion is not inertial because the spaceship is attracted by gravity and therefore is constantly accelerating downward. If you use the spaceship as your reference system, you expect inertial forces, because the spaceship is accelerating. For instance, you should see the centrifugal (apparent) force due to the fact that the orbit is curved downward: masses should accelerate upward.

But they don't.

The reason they don't is that these masses feel the same gravitational pull of the Earth that the spaceship feels. This is an acceleration downward. It is a remarkable fact of gravity that the upward acceleration due to centrifugal force and the downward acceleration due to the Earth's pull *cancel exactly*. So, inside the spaceship things float freely and move in straight lines *as if they were in an inertial system*.

This follows from the fact that all masses fall equally, because m cancels in

$$a = \frac{1}{m} F = \frac{1}{m} \frac{GMm}{r^2} = \frac{GM}{r^2}, \tag{2.4}$$

which shows that the acceleration is independent of m and therefore is the same for all objects: for the spaceship and all the objects within it.

The consequence is spectacular: inside the spaceship, which is falling, the physics is the same as in an inertial system moving uniformly in the absence of gravity.

This means that the effect of gravity on the spaceship can be seen simply as to *redefine the notion of inertial system*: the gravity of the Earth has the effect that the inertial system in which all masses move without accelerating is not the Newtonian one in uniform motion with respect to the fixed stars anymore, but rather the one orbiting the Earth. Thus it is as if gravity determined a new 'true' inertial system.

Now – reasoned Einstein – the role of Newtonian space and time was nothing else than determining the inertial systems. So, both

Newtonian space-and-time and gravity are the 'entity' that locally determines the inertial systems. Hence they must be the same entity.

This spectacularly clever reasoning led Einstein to his most beautiful idea, which is the core of general relativity: *Newtonian space and time and Minkowski spacetime are nothing else than a particular configuration of the gravitational field.* For a general configuration, the geometry of spacetime is not Minkowskian; it is deformed, or 'curved'.

Observation. There is a similarity between this argument by Einstein and the argument used by Newton in the *Principia* to motivate universal gravitation. Newton notes that a body in orbit has the same acceleration as a falling body. He deduces that the cause of the fall and the cause that keeps the celestial bodies in orbit must be the same: universal gravitation. He formalises this argument with what he calls the second 'rule of reasoning': *Causes assigned to effects of the same type must be, as much as possible, the same.* Einstein observes that an inertial local reference frame is determined by the metric structure of spacetime, but also by gravity. He deduces that the metric structure of spacetime and gravity are the same thing: *Causes assigned to effects of the same type must be, as much as possible, the same.*

Newton's bucket, II. So, why does the surface of the water of Newton's bucket become concave? Because the water is in *relative* rotation *with respect to the local gravitational field.* Newton's absolute space is actually the local configuration of the gravitational field. But the gravitational field varies. For instance, as we shall see in Section 10.6, at the North Pole the inertial frame where the water's surface stays flat rotates (slowly) with respect to the fixed stars, because the local gravitational field is affected ('dragged') by the nearby rotating Earth. Newtonian space is replaced by a field, which is affected by the matter of the Earth.

All that remains to do is to learn the maths of curved spaces. This is what happens in the next chapter.

3 Mathematics: Curved Spaces

3.1 CURVED SURFACES

The mathematics that turned out to be effective for describing gravity evolved from that constructed by Carl Friedrich Gauss to describe curved surfaces.[1] I briefly review this theory and the beautiful conceptual discovery by Gauss that opened the door to general relativity.

3.1.1 Intrinsic Geometry

Gauss' brilliant realisation is that there are two distinct ways in which a surface can be curved: it can have 'extrinsic' or 'intrinsic' curvature. Understanding this distinction is the basis for understanding general relativity.

Extrinsic curvature is simple: we say that a 2d surface immersed in 3d Euclidean space has extrinsic curvature if it is not (a portion of) a 2d plane. The definition of *intrinsic* curvature, on the other hand, is Gauss' *stroke of genius*.

- ***Intrinsically flat surfaces***

Imagine a flat sheet of paper, which can bend but not stretch, with some geometrical figures drawn on it (see Figure 3.1, first panel). These obey two-dimensional Euclidean geometry. Imagine bending the paper (Figure 3.1, second panel). When we bend the paper, a straight segment drawn on the paper becomes bent, but it is still the shortest among all the lines between its extremes that can be drawn *on the surface*. Call the shortest line on the surface between two

---

[1] *'Disquisitiones generales circa superficies curvas', auctore Carolo Friderico Gauss, Societati regiae oblate D.8. Octob 1827.*

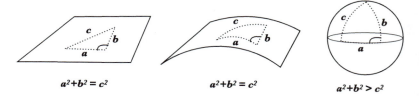

$a^2+b^2 = c^2$ $a^2+b^2 = c^2$ $a^2+b^2 > c^2$

FIGURE 3.1 Left: a triangle on a plane surface. Centre: a triangle on a bent surface. Right: a triangle on a sphere. Both the surface in the central panel and the sphere are *extrinsically* curved, but the first is *intrinsically flat*, the second is *intrinsically curved*.

given points '(intrinsically) straight', and call its length the '(intrinsic) distance' between its extremes. Obviously these '(intrinsically) straight lines' and the '(intrinsic) distances' between the points on the bent sheet satisfy the same properties as the straight lines and the distances on a 2d plane.

For instance, imagine a triangle drawn on the paper. Neither the length of its sides nor the amplitude of its angles changes when bending the paper. Therefore, the triangle drawn on the bent paper satisfies the standard properties of 2d Euclidean triangles: if one angle is straight, the lengths a, b, c of its sides satisfy Pythagoras' theorem $a^2 + b^2 = c^2$; the sum of the three angles is π(rad). Similarly, a circle drawn on the paper (the set of points at equal intrinsic distance r from a central point) has a perimeter p that still satisfies $p = 2\pi r$ when drawn on the paper. And so on: standard Euclidean geometry holds.

Let's put it visually: if you were a small ant moving on the bent paper and capable of measuring angles and lengths of lines on the surface, but incapable of looking 'outside' the paper, you would not be able to figure out that you are on a surface that is not a plane. The two-dimensional geometry defined by the length of the intrinsically straight lines is the same geometry as the geometry of a plane. This geometry is called 'intrinsic geometry'. Hence we say that 'the intrinsic geometry of the bent sheet of paper is flat', even if the paper itself is actually curved.

• Intrinsically curved surfaces

What is described above, however, is not true for generic curved surfaces. Consider, for instance, a sphere of unit radius. The intrinsic geometry defined by the length of the lines drawn on the sphere itself is *not* the geometry of a plane.

Given two points on the sphere, the shortest line on the sphere connecting them is a portion of a maximal circle. These are the 'intrinsically straight' segments on the sphere. Take the North Pole of the sphere and two points on the equator at one quarter of the equator length from one another. These define a triangle, formed by a portion of the equator and two meridians. It is evident immediately that the sum of the angles of this triangle is not π but $\frac{3}{2}\pi$! Similarly, the equator is a circle or length $p = 2\pi$ at intrinsic distance $r = \pi/2$ from the North Pole, hence it does not satisfy $p = 2\pi r$, but rather $p = 4r$.

Thus, straight lines on the sphere *define an intrinsic geometry which is different from the geometry of the 2d plane*. If you were a small ant moving on the sphere, capable of measuring lengths of lines on the surface but incapable of looking 'outside' the surface, you *would* be able to figure out that you are not on a plane: it would suffice to measure the length p of the line formed by points at a distance r from a centre: if $p \neq 2\pi r$, your intrinsic geometry is not flat. When the intrinsic geometry is not flat, we say that the surface has 'intrinsic curvature'.

These examples illustrate the difference between extrinsic and intrinsic curvature.

• Intrinsic geometry

The 'intrinsic geometry' of a surface is the geometry defined by the length of the lines lying on the surface. If this geometry is the same as that of the lines on a plane, we say that the surface is 'intrinsically flat', or that it has no 'intrinsic curvature'. If instead, the geometry defined by these lines is different from the geometry of the lines on a plane, we say that the geometry is 'intrinsically curved', or that there is 'intrinsic curvature'.

The importance of this intuition by Gauss is that in this manner we can talk about the curvature of a surface *using only the geometry of the distances on the surface itself, with no need of looking at how the surface is embedded in a larger space.* This is exactly what we shall do in general relativity.

3.1.2   Gauss Curvature

Let us make the above definitions quantitative. Consider a closed path on the surface starting and ending at p, composed of straight lines and encircling a surface with small area A. Imagine to parallel transport of a vector along this curve, namely to carry it along the path, keeping the angle with each straight path constant. If the surface is flat, the vector will return oriented as it started. In general, it will be rotated by an angle α. The Gauss curvature at p is

$$K = \lim_{A \to 0} \frac{\alpha}{A}. \tag{3.1}$$

A sphere is uniform, so its curvature is the same at each point. Its curvature is

$$K = \frac{1}{R^2}, \tag{3.2}$$

where R is the radius. *[Compute it. A simple path to use is a triangle with one vertex on a pole and two on the equator. See Figure 3.2. Bring these two vertices close by.]* On a general surface, on the other hand, such as the flying object in front of the Sistine Chapel frescos on the cover of this book, or the surface of the comet of Figure 3.3, the curvature changes from point to point, defining a field $K(x)$ on the surface.

An equivalent definition of curvature is the following. Consider a small circle of radius r and perimeter P centred at a point p on the surface: define the 'Gauss curvature' K of the surface at p by

$$K = \frac{3}{\pi} \lim_{r \to 0} \frac{2\pi r - P}{r^3}. \tag{3.3}$$

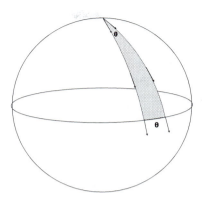

FIGURE 3.2 The definition of curvature via parallel transport. The triangle's area is $A = \theta R^2$ (easily from a proportion with the area of the hemisphere). The vector returns rotated by an angle $\alpha = \theta$.

[Compute K for a sphere, using this formula.]

Gauss' definition. In his founding work on differential geometry, cited at the beginning of this chapter, Gauss does not give an intrinsic definition of Gauss curvature such as those given above. He gives a definition that uses the embedding, and then he proves that it is independent from the embedding. Gauss' definition is extremely beautiful: every point p of a surface Σ immersed in R^3 determines a direction $\vec{n}_p$ perpendicular to Σ at p; consider the sphere of unit radius of all possible directions; given a region $R \subset \Sigma$, the set of directions $\vec{n}_p$ for all $p \in R$ draws a region R_0 of this sphere. The curvature of Σ at p is defined as the limit for a small area of the ratio of the areas of R_0 and R. Gauss calls the theorem stating that this curvature is independent from the immersion *Teorema Egregium*, which means 'outstanding theorem'. This theorem is the conceptual basis of the mathematics of general relativity: curvature is a concept independent from the embedding.

- *Curvature sees the second order in the distance*

Consider the neighbourhood of a point x on the surface. Up to *first* order in the distance, the geometry of the surface is approximated by the plane tangent to the surface. This is the geometrical definition of tangent plane, in fact. Intuitively, a (smooth) curved surface always 'looks flat' if we restrict ourselves to a sufficiently small region. This is familiar: the surface of the Earth is spherical but can be approximated by a plane in sufficiently small regions. This is why a plane map of a small Earth region can reproduce the geometry of the Earth sufficiently faithfully. The Gauss curvature is the quantity that measures the scale at which this approximation fails.

FIGURE 3.3 A curved surface: the surface of the comet on which the Rosetta space probe has landed. There are no 'natural' coordinates for this surface. (ESA/Rosetta/NavCam; CC BY-SA 3.0 IGO)

To *second* order in the distance, at a point x where $K(x) > 0$, the geometry of the surface is approximated by the geometry of a sphere of radius R, related to the curvature by (3.2). Therefore $K(x)$ is (the inverse of the square of) the radius of the sphere that best approximates the surface at x. If $K(x) < 0$, the surface is approximated by the hyperboloid of radius R (namely the surface defined by $x^2 + y^2 - z^2 = R^2$).

If the curvature is zero, we say that the surface is '(intrinsically) flat'. A cylinder is intrinsically flat.

3.1.3   General Coordinates

The tool introduced by Gauss to describe curved surfaces is general coordinates. Consider a smooth (C^∞) two-dimensional surface Σ, possibly curved, immersed in the familiar three-dimensional Euclidean space. Let $X^I = (X^1, X^2, X^3) = (X, Y, Z)$ be the Cartesian coordinates of the three-dimensional space. To describe a generic surface Σ, we can proceed as follows. We choose *arbitrary* coordinates $x^a = (x^1, x^2)$ on the surface and we specify the surface by giving the location of the point with coordinates x^a in the Euclidean space. That is, we give the three functions of two variables

$$\Sigma : x^a \mapsto X^I(x^a). \tag{3.4}$$

For instance, a sphere S_2 of unit radius can be coordinated by polar coordinates $x^a = (\theta, \phi)$, with $\theta \in [0, \pi]$, $\phi \in [0, 2\pi]$ and defined by the well-known functions

$$X = \sin \theta \cos \phi, \quad Y = \sin \theta \sin \phi, \quad Z = \cos \theta. \tag{3.5}$$

The x^a coordinates are arbitrary: we can choose them differently in innumerable other manners. For instance, on the same sphere we can choose the coordinates $\tilde{x}^a = (z, \phi)$ with $z \in [-1, 1]$ related to the previous ones by

$$z = \cos \theta. \tag{3.6}$$

In terms of these coordinates, the equations describing the sphere are

$$X = \sqrt{1 - z^2} \cos \phi, \quad Y = \sqrt{1 - z^2} \sin \phi, \quad Z = z. \tag{3.7}$$

In general, assigned a surface given in coordinates x^a, we can always rewrite it in terms of new arbitrary coordinates $\tilde{x}^a$ related to the previous ones by any smooth and invertible function

$$\tilde{x}^a = \tilde{x}^a(x^a). \tag{3.8}$$

Why are we using *arbitrary* coordinates? Because on a general curved surface nothing selects preferred coordinates, or a preferred family of coordinates, in the manner Cartesian coordinates are preferred on a Euclidean space. If the surface has some kind of symmetry (like a sphere), it is convenient to adopt coordinates adapted to this symmetry (like the polar coordinates above). But there are no 'natural' coordinates on an arbitrary surface like the surface of the bumpy comet in Figure 3.3. To be able to describe arbitrary curved spaces, we have to live with arbitrary coordinates.

As we shall see, this freedom of choosing coordinates in describing curved spaces plays a major role. It has also historically been a source of much confusion in general relativity.

• Difference between Euclidean and general coordinates

On a Euclidean space there are (families of) natural coordinates: the Cartesian coordinates. They express *geometrical distances* from the planes of an orthogonal reference system. On an arbitrary curved surface, there are no such preferred coordinates, in general. The arbitrary coordinates used to coordinate a curved surface *do not have the meaning of distances*.

Einstein wrote that his greatest difficulty in constructing general relativity was his struggle to 'understand the meaning of the coordinates'. The conceptual difficulty was to separate the notion of coordinate from the notion of distance.

Coordinate singularities. A further feature that complicates the use of general coordinates is that it is often impossible (or inconvenient) to have coordinates that cover the full surface. Something may go wrong somewhere.

For instance, the familiar polar coordinates (θ, ϕ) of a sphere behave funnily at the poles: all the points $(0, \phi)$ for different values of ϕ are actually the same point!

Here is another example that plays an important role below, for black holes. Consider a very simple surface: the plane. We can coordinate it with Cartesian coordinates X, Y both $\in [-\infty, \infty]$. These are 'good' coordinates that cover the full plane. But let us choose instead coordinates $x = X, y = Y - \frac{1}{X}$. The coordinates x, y are ill-behaved at $x = 0$. In fact, the lines of equal y are drawn in Figure 3.4 in the plane of the Cartesian coordinates Y, X, which makes it clear that they go bad at $X = 0$. What precisely happens is that the line $X = 0$ is only reached for $y \to \infty$, which means that it is never reached for any finite y. It is as if these coordinates

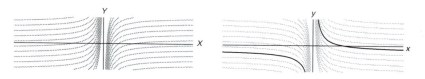

FIGURE 3.4 Left: the lines of constant y coordinates on the plane, in the Cartesian coordinates X, Y. Right: the lines of the constant Cartesian coordinate Y in the plane of the non-Cartesian coordinates x, y. The dark line is the Cartesian axis $Y = 0$; notice that it reaches $x = 0$ only at infinity: coming from positive x, something strange appears to happen at $x = 0$, but this is only the effect of a 'bad' coordination of a very regular Cartesian plane.

avoid the $X = 0$ line. In a sense, this line is thrown out of the space and sent to infinity if we use these 'bad' coordinates.

The moral is that in using arbitrary coordinates, one should always be careful that some apparently strange phenomena might simply be 'coordinate artefacts': obviously there is nothing strange in the geometry of the North Pole or the y-axis of a plane. As we shall see, this is a complication that for several decades confused all physicists, including Einstein, on the nature of the black holes.

I now introduce two fundamental quantities that describe the intrinsic geometry of the surface without making use of the embedding.

3.1.4   Frame Field and Metric

• **Frame field**

Since the surface Σ is smooth, at any point p we can approximate it by the plane tangent at p. Consider a system of *Cartesian* coordinates $X_p^i = (X_p, Y_p)$ on this tangent plane, with origin at p. If we project these coordinates down to the surface (orthogonally to the plane), we obtain local coordinates $X_p^i = (X_p, Y_p)$ on a neighbourhood of p, which are called 'local Cartesian coordinates at p'. Given arbitrary general coordinates x^a on Σ, let x_p^a be the coordinates of the point p, and consider the map $X_p^i(x^a)$ from these arbitrary coordinates to the local Cartesian coordinates. The Jacobian of this map at p is the 2×2 matrix

$$e_a^i = \left. \frac{\partial X_p^i(x^a)}{\partial x^a} \right|^{x_a = x_p^a} . \tag{3.9}$$

We can repeat this procedure at each point p of Σ with coordinates x^a thus obtaining a *field* $e_a^i(x^a)$ on the surface. This field is called a 'frame field', or a 'diad' field. It expresses the relation between the arbitrary coordinates x^a at a point and Cartesian coordinates X^i at this point. As we shall see, this field describes gravity.

• **Inverse frame field**

If the quantities x^a are good coordinates around p, the determinant of the Jacobian, and therefore the determinant of e_a^i is non-vanishing.

We can therefore consider the inverse of this 2×2 matrix. The inverse is denoted e_i^a (indices swapped) and is, of course, the Jacobian of the inverse transformation: from the arbitrary coordinates to the Cartesian coordinates

$$e_i^a = \frac{\partial x^a(X_p^i)}{\partial X_p^i}\bigg|_{X_p^i=0}. \tag{3.10}$$

It carries the same information as the frame field. This object can be seen as a couple of vector fields tangent to the surface:

$$e_i^a(x) = (\vec{e}_1(x), \vec{e}_2(x)). \tag{3.11}$$

At each point the two vectors form an orthonormal basis, because they point in the directions of the local Cartesian coordinates. See Figure 3.5. The name 'frame field' comes from this fact: at each point, the field defines a local Cartesian reference frame.

Here is a concrete example. Except for the poles, on a small neighbourhood of a generic point with coordinates (θ, ϕ) of a unit sphere, the coordinates $X(\theta', \phi') = \theta' - \theta$ and $Y(\theta', \phi') = \sin\theta(\phi' - \phi)$ are clearly local Cartesian coordinates. Therefore, a frame field for the unit sphere is given by

$$e_a^i(\theta, \phi) = \begin{pmatrix} \frac{\partial X}{\partial \theta'}\big|_{\theta,\phi} & \frac{\partial X}{\partial \phi'}\big|_{\theta,\phi} \\ \frac{\partial Y}{\partial \theta'}\big|_{\theta,\phi} & \frac{\partial Y}{\partial \phi'}\big|_{\theta,\phi} \end{pmatrix} = \begin{pmatrix} 1 & 0 \\ 0 & \sin\theta \end{pmatrix}. \tag{3.12}$$

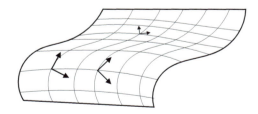

FIGURE 3.5 A frame field $e_i^a(x) = (\vec{e}_1(x), \vec{e}_2(x))$ is a choice of two orthonormal tangent vectors at each point of the surface. Knowing these fields on the surface is sufficient to determine its intrinsic geometry, without knowing how the surface is embedded in a three-dimensional space.

(Do not confuse the coordinates θ, ϕ where the field is calculated with the coordinates θ', ϕ' on a neighbourhood of this point.)

- $SO(2)$ *gauge*

In the definition of the frame there is some freedom, because Cartesian coordinates are not unique. At each point p with coordinates x^a we can choose different Cartesian coordinates $\tilde{X}^i$. Two sets of Cartesian coordinates are related by $\tilde{X}^i = R^i_j X^j$, where R is a rotation matrix in $SO(2)$. Clearly under a change of frame the diad changes as $e^i_a \rightarrow R^i_j e^j_a$, and since we can choose Cartesian systems arbitrarily at each point, the entire frame field can change as

$$e^i_a(x) \rightarrow R^i_j(x)e^j_a(x).$$ (3.13)

This is called an $SO(2)$ gauge transformation of the frame field.

- *The frame field captures the intrinsic geometry*

Now I come to the main point. Consider a point on the surface with coordinates x^a, and a nearby point with coordinates $x^a + dx^a$. What is the *distance* between these two points?

If the points are close, we can work to first order in dx^a. The change in Cartesian coordinates between the two points is

$$dX^i = \frac{\partial X^i(x^a)}{\partial x^a} \, dx^a = e^i_a(x)dx^a$$ (3.14)

and their distance ds is given in Cartesian coordinates by

$$ds^2 = \sum_i dX^i dX^i \equiv \delta_{ij} \, dX^i dX^j,$$ (3.15)

where δ_{ij} is the unit matrix, namely $\delta_{ii} = 1$ and $\delta_{ij} = 0$ if $i \neq j$ and sum over repeated indices is understood. Bringing the last two equations together we have

$$ds^2 = \delta_{ij}e^i_a(x)e^j_b(x)dx^a dx^b.$$ (3.16)

Therefore, if we know the frame field $e^i_a(x)$, we can calculate the distance between points on the surface. The frame field determines intrinsic distances on the surface: it fully characterises the intrinsic geometry of the surface.

- **Metric**

It is convenient to write (3.16) in the form

$$ds^2 = g_{ab}(x)dx^a dx^b \qquad (3.17)$$

where

$$g_{ab}(x) = e^i_a(x)e^j_b(x)\delta_{ij}. \qquad (3.18)$$

The field $g_{ab}(x)$ is called the Riemann metric, or simply the 'metric' of the surface Σ. It is manifestly symmetric $(g_{ab} = g_{ba})$, by construction has non-vanishing determinant, and all its eigenvalues are positive (this can be easily seen by diagonalising the diad). It is easy to show [*do it!*] that it is invariant under the $SO(2)$ gauge transformation (3.13), because every rotation metric R satisfies $R^i_k R^l_m \delta_{il} = \delta_{km}$.

As we shall see, $g_{ab}(x)$ is the field that Einstein used to express the gravitational field. It later turned out that $g_{ab}(x)$ is not sufficient to describe gravity in general (for instance, in the presence of fermions) and the frame field $e^i_a(x)$ gives a more complete description.

The interest of the metric field, or the frame field, is that they capture the intrinsic geometry of the surface, without any reference to the way the surface is embedded in the three-dimensional Euclidean space (we shall soon see why this is crucial for physics). This can be seen explicitly as follows.

- **Length of a curve**

Remember that the intrinsic geometry is the geometry defined by the *length* of the curves on the surface. Given a curve γ on the surface, we can compute its length using the metric as follows. Let τ be an arbitrary parameter along the curve, so that the curve is expressed by the two functions of one variable

$$\gamma : \tau \mapsto x^a(\tau). \qquad (3.19)$$

An infinitesimal displacement $d\tau$ along the curve gives rise to a coordinate change $dx^a = \frac{dx^a}{d\tau}d\tau \equiv \dot{x}^a d\tau$ whose length, using (3.17), is

$$ds = \sqrt{g_{ab}(x(\tau))dx^a dx^b} = \sqrt{g_{ab}(x(\tau))\dot{x}^a\ d\tau\ \dot{x}^b\ d\tau}$$

$$= \sqrt{g_{ab}(x(\tau))\dot{x}^a\dot{x}^b}\ d\tau. \qquad (3.20)$$

The length of the curve is then obtained by integrating this expression along the curve

$$L[\gamma] = \int \sqrt{g_{ab}(x(\tau))\dot{x}^a\dot{x}^b}\ d\tau \equiv \int \sqrt{g_{ab}\dot{x}^a\dot{x}^b}\ d\tau. \qquad (3.21)$$

This expression is invariant under changes of coordinates on the surface [show it!] as well as under change of parametrisation of the curve [show it!]. It depends only on the curve and the geometry of the surface, not on the coordinates that we use on the surface or on the curve itself.

If we know the metric field $g_{ab}(x)$ of a surface, we know the length of any curve on the surface. By the very definition of intrinsic geometry, then, the metric field $g_{ab}(x)$, or the diad field $e_a^i(x)$ from which the metric can be computed, captures the intrinsic geometry of the surface entirely.

Since the Gauss curvature $K(x)$ depends only on the local intrinsic geometry, and since this is entirely determined by the metric, it must be possible to express $K(x)$ as a local function of $g_{ab}(x)$ and its (first and second) derivatives. This is indeed the case. I do not give here the explicit formula, because it will be simpler to give it in the next chapter, in the general case of spaces of arbitrary dimensions, which we will see shortly.

• *Angles*

The metric allows us to compute angles between vectors on the surface as well. A vector $\vec{V}$ at a point x^a of the surface is a vector in the tangent space at x^a. If its components with respect to the Cartesian coordinates X^i are V^i, its components with respect to the general coordinates x^a are defined by

$$V^a \equiv e^a_i(x) V^i. \tag{3.22}$$

The angle θ between two vectors of unit length in the tangent space at a point p with components V^i and W^i with respect to the Cartesian coordinates X^i is given by elementary geometry:

$$\cos\theta = \vec{V} \cdot \vec{W} = \delta_{ij} V^i W^j. \tag{3.23}$$

Using the definitions of the metric and V^a this gives

$$\cos\theta = g_{ab}(x) V^a W^b. \tag{3.24}$$

And the length $|V|^2 = \delta_{ij} V^i V^j$ of a vector is

$$|V|^2 = g_{ab}(x) V^a V^b. \tag{3.25}$$

- **_Inverse frame field and inverse metric_**

Using the inverse frame field, the definition of the metric becomes

$$e^a_i(x) e^b_j(x) g_{ab}(x) = \delta_{ij}, \tag{3.26}$$

which shows explicitly that the inverse frame field represents two vector fields $e^a_i(x) = (\vec{e}_1(x), \vec{e}_2(x))$ that at each point define an orthonormal frame in the tangent space as in Figure 3.5. That is,

$$\vec{e}_i(x) \cdot \vec{e}_j(x) = \delta_{ij}. \tag{3.27}$$

If we invert the 2×2 matrix g_{ab}, we obtain a matrix that is called g^{ab} (sometimes 'contravariant metric'), and has the same information as g_{ab}. Using this we can write, from the last equation,

$$g^{ab}(x) e^i_a(x) \delta_{ij} = e^b_j(x). \tag{3.28}$$

This shows that the 'space' indices $a, b \ldots$ can be consistently raised and lowered with the metric and its inverse, while the 'internal' indices $i, j \ldots$ can be consistently raised and lowered with the Kronecker delta δ_{ij}.

Notice that the metric information about the surface (lengths, angles...) is in the metric field (or the frame field), not in the coordinates. The Cartesian coordinates X^i are defined in such a way that they immediately give distances, but not so for general coordinates x^a: these have no metric information.

- **Coordinate change and invariant geometry**

If we change coordinates on the surface, it is an easy exercise *[do it!]* to show that the frame field changes as follows:

$$e^i_a(x) \to \tilde{e}^i_a(\tilde{x}) = \frac{\partial x^b}{\partial \tilde{x}^a}\, e^j_b(x(\tilde{x})), \tag{3.29}$$

and the metric field changes as follows:

$$g_{ab}(x) \to \tilde{g}_{cd}(\tilde{x}) = \frac{\partial x^a}{\partial \tilde{x}^c}\frac{\partial x^b}{\partial \tilde{x}^d}\, g_{ab}(x(\tilde{x})). \tag{3.30}$$

Two metrics related by this transformation describe the same surface, in different coordinates. Therefore what matters for the geometry of the surface is not the field $g_{ab}(x)$, but rather the equivalence class of these fields under the equivalence defined by (3.30). These equivalent classes are called two-dimensional *geometries*. For instance, a (metric) sphere is a two-dimensional geometry and can be described by different $g_{ab}(x)$.

3.2 RIEMANNIAN GEOMETRY

In 1853, Gauss gave a student the thesis ('Habilitationsschrift') topic of generalising the above construction to higher dimensions by defining and studying 'curved spaces of higher dimensions'. The name of the student was Bernhard Riemann. The result was what we today call 'Riemannian geometry'.[2] This is the mathematics used by Einstein to describe gravity.

Riemann's idea was to define *only* the *intrinsic* geometry of a space. If we have a space coordinated by coordinates x^a and we assign the metric field $g_{ab}(x)$ or the diad field $e^i_a(x)$, then we know the intrinsic geometry of this space, because this is sufficient to compute the length of any curve from equation (3.17). We do not need to

---

[2] B. Riemann, *Über die Hypothesen, welche der Geometrie zu Grunde liegen*, 'On the hypothesis on which geometry is based', Abhandlungen der Königlichen Gesellschaft der Wissenschaften zu Göttingen, vol. 13, 1867.

know how it is embedded in a higher-dimensional space, because this information has no relevance for its intrinsic geometry. We can therefore consider just the space with its intrinsic geometry, without any embedding in a higher-dimensional space. Once this is understood, it is easy to generalise Gauss' construction to higher-dimensional spaces.

- **Definition of Riemannian space**

Consider a space of dimension d on which general coordinates $x^a = (x^1, x^2, \ldots, x^d)$ are defined, and a frame field $e^i_a(x)$, where the index i runs from 1 to d as well. (For $d = 3$ this is also called a 'triad', and for $d = 4$ a 'tetrad'. The frame field is also denoted by the German expression 'd-bein' field.) The metric field is immediately defined by (3.18). The intrinsic distance ds between two points with coordinates x^a and $x^a + dx^a$ is *defined* by (3.17), which I repeat here, since it is the basic equation of Riemannian geometry: it defines the geometry from the Riemann metric

$$\boxed{ds^2 = g_{ab}(x)\, dx^a dx^b}.$$

(3.31)

The coordinates and the metric define a Riemann space. The length of a curve is defined by (3.21). The intrinsic geometry is defined by this notion of length, without reference to an embedding of the space in higher dimensions. Since coordinates can be changed arbitrarily, a Riemannian geometry is an equivalence class of metric fields $g_{ab}(x)$ under the equivalence relation (3.30).

- **Flat and curved Riemann geometries**

A d-dimensional Riemannian space is said to be 'flat' if the intrinsic geometry it defines is the same as that of a d-dimensional Euclidean space. In this case we say that its curvature is zero. It is clear that in this case we can always find Cartesian coordinates in which the frame field and the metric have the form

$$e^i_a(x) = \delta^i_a, \qquad g_{ab}(x) = \delta_{ab}.$$

(3.32)

In other words, a flat Riemannian space is the same thing as a d-dimensional Euclidean space. But in general a Riemannian space is not flat: it is a metric space where Pythagoras' theorem does not hold, the sum of the angles of triangles can be different from π(rad), and so on, as in the intrinsic geometry of the curved surfaces studied by Gauss, but without a need to embed the space in higher dimensions.

As we shall see, the *physical* space in which we actually live is a curved Riemann space: Pythagoras' theorem does not hold in our universe. It is violated slightly (because the curvature is small), but it is violated.

Of course, it is not the mathematical theorem that is wrong: it is its hypotheses, namely the Euclidean postulates, that are not satisfied by the real physical space.

• *Examples of Riemannian spaces*

To specify a Riemannian space, we have to specify the coordinates (with their range, if this is not understood) and give the field $e^i_a(x)$, or the metric field $g_{ab}(x)$, as functions of these coordinates. A convenient and common way of giving $g_{ab}(x)$ is to write ds^2 explicitly. Here are some examples

$$\text{Plane} : ds^2 = dx^2 + dy^2. \tag{3.33}$$

$$\text{Unit sphere} : ds^2 = d\theta^2 + \sin^2\theta \, d\phi^2 \equiv d\Omega^2. \tag{3.34}$$

$$\text{Ellipsoid} : ds^2 = a \, d\theta^2 + b \sin^2\theta \, d\phi^2. \tag{3.35}$$

$$\text{3d Euclidean space} : ds^2 = dx^2 + dy^2 + dz^2. \tag{3.36}$$

The first line means $g_{ab}(x, y) = \begin{pmatrix} 1 & 0 \\ 0 & 1 \end{pmatrix}$, the second $g_{ab}(\theta, \phi) = \begin{pmatrix} 1 & 0 \\ 0 & \sin^2\theta \end{pmatrix}$, and so on. The quantity ds written in this form is called the 'line element'. An advantage of this notation is that changing coordinates is particularly simple. For instance, suppose we want to go to polar coordinates on the plane:

$$x = r \sin\phi, \quad y = r \cos\phi. \tag{3.37}$$

Then we can find the metric of the plane in the coordinates r, ϕ by simply differentiating (3.37)

$$dx = dr \sin\phi + r \cos\phi \, d\phi, \quad dy = dr \cos\phi - r \sin\phi \, d\phi \tag{3.38}$$

and replacing this *directly* in (3.33), which gives

$$\text{Plane in polar coordinates} : \quad ds^2 = dr^2 + r^2 \, d\phi^2. \tag{3.39}$$

- **Derivation of the metric of the 2-sphere from the embedding**

A way to derive the metric (3.34) of a sphere is to start by viewing the sphere as the surface defined by

$$X^2 + Y^2 + Z^2 = 1 \qquad (3.40)$$

in the three-dimensional Euclidean space with metric

$$ds^2 = dX^2 + dY^2 + dZ^2. \qquad (3.41)$$

It is convenient to go to polar coordinates r, ϕ in the X, Y space, which gives

$$ds^2 = dr^2 + r^2 d\phi^2 + dZ^2, \qquad (3.42)$$

and the 2-sphere of unit radius is defined by $r^2 + Z^2 = 1$. Differentiating this relation gives $2Z dZ = -2r dr$. Hence $dZ = -r dr / \sqrt{1 - r^2}$, so that

$$ds^2 = dr^2 + r^2 d\phi^2 + \frac{r^2 dr^2}{1 - r^2} = \frac{dr^2}{1 - r^2} + r^2 d\phi^2, \qquad (3.43)$$

where $r \in [0, 1]$. If we change coordinates introducing $r = \sin\theta$ [do it!], we obtain (3.34).

    Notice that the coordinates (r, θ, ϕ) in (3.43) go bad on the equator $r = 1$, where the component $g_{rr}(r, \phi)$ of the metric becomes infinite. This is an example of 'coordinate singularity': a divergence of the metric that reflects a bad behaviour of the coordinates, not anything peculiar happening to the actual geometry of the space. The problem can be cured simply by changing coordinates (here: $r \to \theta$).

- **3-sphere**

The conventional sphere, also called the 2-sphere, is a homogeneous (all points have the same properties) and isotropic (all directions at any point have the same properties) 2d space with finite volume and no boundaries. The homogeneous and isotropic 3d space with finite volume and no boundaries is called '3-sphere'.

    To write the metric of the 3-sphere, we can follow the same procedure used above to derive the metric of the 2-sphere, but in one dimension above. The 3-sphere is the surface defined by

$$X^2 + Y^2 + Z^2 + U^2 = 1 \qquad (3.44)$$

in a four-dimensional Euclidean space with metric

$$ds^2 = dX^2 + dY^2 + dZ^2 + dU^2. \qquad (3.45)$$

It is convenient to go to polar coordinates r, θ, ϕ in the X, Y, Z space, which gives

$$ds^2 = dr^2 + r^2 d\Omega^2 + dU^2, \qquad (3.46)$$

$(d\Omega^2$ is defined in (3.34)) and $r^2 + U^2 = 1$. Differentiating this gives $2UdU = -2rdr$. Hence $dU = -rdr/\sqrt{1-r^2}$, so that

$$ds^2 = dr^2 + r^2 d\Omega^2 + \frac{r^2 dr^2}{1-r^2} = \frac{dr^2}{1-r^2} + r^2 d\Omega^2. \qquad (3.47)$$

This is the metric of the unit 3-sphere in r, θ, ϕ coordinates. Explicitly:

$$\text{Unit 3-sphere:} \quad ds^2 = \frac{dr^2}{1-r^2} + r^2(d\theta^2 + \sin^2\theta \, d\phi^2). \qquad (3.48)$$

In terms of all angular coordinates, using $r = \sin\psi$, this becomes

$$\text{Unit 3-sphere:} \quad ds^2 = d\psi^2 + \sin^2\psi \, (d\theta^2 + \sin^2\theta \, d\phi^2), \qquad (3.49)$$

where $\psi \in [0, \pi], \theta \in [0, \pi], \phi \in [0, 2\pi]$. The metric of a larger sphere is simply obtained by multiplying the components of the metric tensor by a positive constant a^2. That is

$$\text{3-sphere:} \quad ds^2 = a^2 \left(\frac{dr^2}{1-r^2} + r^2(d\theta^2 + \sin^2\theta \, d\phi^2) \right). \qquad (3.50)$$

The quantity a is called the radius.

One way to visualise the conventional 2-sphere is by drawing two equal discs (the northern and southern hemispheres) glued by their common boundary (the equator: a circle). One way to visualise a 3-sphere is to imagine two equal balls (the northern and southern hemispheres) glued by their common boundary (the equator: a 2-sphere). See Figure 3.6.

Exercise: *Write the metric of the 4-sphere.*

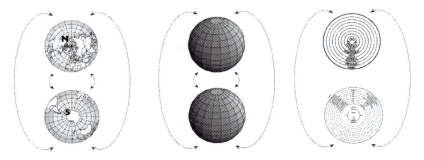

FIGURE 3.6 Left: A 2-sphere is obtained by gluing two discs by their boundary (a circle). Centre: A 3-sphere is obtained by gluing two balls by their boundary (a 2-sphere). Right: The universe described by Dante is a 3-sphere.

- **Homogeneous spaces**

The 3-sphere is a homogeneous isotropic space with constant *positive* curvature. A homogeneous isotropic space with *zero* curvature is, of course, Euclidean space. There is also a homogeneous isotropic space with constant *negative* curvature: a Lorentz hyperboloid in 4-dimensional Minkowski space. Repeating the same steps as above gives

$$\text{Three-hyperboloid}: \quad ds^2 = a^2 \left(\frac{dr^2}{1+r^2} + r^2(d\theta^2 + \sin^2\theta \, d\phi^2) \right).$$

(3.51)

The metric

$$\text{3d homogeneous space}: \quad ds^2 = a^2 \left(\frac{dr^2}{1 - kr^2} \right.$$
$$\left. + r^2(d\theta^2 + \sin^2\theta \, d\phi^2) \right)$$

(3.52)

where k can take the value 1 (sphere), -1 (hyperboloid), or 0 (Euclidean space), can be shown to be the most general homogeneous isotropic 3d metric. These metric spaces play a fundamental role in modern relativistic cosmology, as we shall see later on.

- **Dante Alighieri**

As far as I know, the first visualisation of a 3-sphere can be found in the medieval poem the *Divina Commedia* by the great Italian poet, Dante Alighieri. Dante depicts the universe as formed by two balls glued by their boundary: the first is centred on the Earth and bounded by the sphere of the fixed stars, the other has God at the centre with spheres of angels surrounding Him and it is also bounded by the same sphere of the fixed stars. In Dante's words, the two balls 'surround each other', as is precisely the case for a 3-sphere. See Figure 3.6.

- **Brunetto Latini**

It is surprising that Dante could have had the intuition of the existence of this mathematical object in the fourteenth century, but two considerations make this fact a bit more plausible. The first is that before Newton, the idea of an infinite space with the metric structure of Euclidean geometry was not yet current (remember that Euclid described the geometry of ideal figures, not space).

The second is more interesting, in relation to the mathematics I am describing. Dante's main sources were the books of his teacher, Brunetto Latini. In these, the spherical shape of the surface of the Earth is accurately described but – surprisingly for a model reader – not in extrinsic terms, but in intrinsic terms. That is, Brunetto does not write 'The Earth is like an orange'. He writes: 'If a knight rides in a direction and if nothing stopped him, he would get back to the departing point'. The reason for this curious way of describing a sphere is – I suspect – to emphasise the homogeneity of the physics on the surface of the Earth: an orange always has an 'upper side' where things fall towards it and a 'lower side' where things fall away from it. Not so the Earth, as Brunetto knew well. Now, if young Dante learned the geometry of a sphere in this intrinsic manner, it is less surprising that he could have imagined the intrinsic geometry of a 3-sphere. After all, a 3-sphere is simply a space where 'If a spaceship flies in a direction and if nothing stopped it, it would get back to the departing point'.

3.2.1 Geodesics

The shortest lines between two given points, namely the 'intrinsically straight' lines, are called 'geodesics'.

In Riemannian geometry they play for the geometry the same role straight lines play in Euclidean geometry. They are defined by the fact that they minimise (3.21) among all the curves with the same extremes.

- **The geodesic equation**

Geodesics satisfy a local equation called the geodesic equation, which plays a central role in general relativity. Let us derive it explicitly. If we vary the curve in (3.21) (keeping the extreme fixed), its length changes as follows:

$$\delta L[\gamma] = \int \delta \sqrt{g_{ab}\dot{x}^a\dot{x}^b}\, d\tau = \int \frac{\delta g_{ab}\dot{x}^a\dot{x}^b + 2g_{ab}\delta\dot{x}^a\dot{x}^b}{2\sqrt{g_{ab}\dot{x}^a\dot{x}^b}} d\tau. \quad (3.53)$$

It is convenient to parametrise the curve γ with a parameter τ equal to the length, that is, to pose $d\tau = ds$. In this parametrisation

$$|\dot{x}| = \sqrt{g_{ab}\dot{x}^a\dot{x}^b} = 1, \quad (3.54)$$

and this makes the square root in the denominator equal to unit. (We could not have fixed this parametrisation *before* doing the variation and also kept the condition that the variation δx^c vanishes at the boundaries.) Since first-order variations commute, $\delta\dot{x}^a = d/d\tau(\delta x^a)$, we can integrate the τ derivative by parts, obtaining

$$\delta L[\gamma] = \frac{1}{2}\int \left(\partial_c g_{ab}\dot{x}^a\dot{x}^b - 2\frac{d}{d\tau}(g_{cb}\dot{x}^b) \right) \delta x^c d\tau, \quad (3.55)$$

where $\partial_c g_{ab} \equiv \frac{\partial g_{ab}}{\partial x^c}$. The boundary terms of the integration by parts vanish because we have assumed that the variation δx^c of the curve vanishes at the extremes. If the original curve is a geodesic, namely a minimal-length curve, this variation must vanish for any $\delta x^c(\tau)$, hence the parentheses must vanish. This gives

$$\partial_c g_{ab} \dot{x}^a \dot{x}^b - 2\partial_a g_{cb} \dot{x}^a \dot{x}^b - 2g_{cb} \ddot{x}^b = 0. \tag{3.56}$$

It is convenient to split the second term in two, exploiting the symmetry of $\dot{x}^a \dot{x}^b$, and to bring the last term on the other side of the equality

$$\partial_c g_{ab} \dot{x}^a \dot{x}^b - \partial_a g_{cb} \dot{x}^a \dot{x}^b - \partial_b g_{ca} \dot{x}^a \dot{x}^b = 2g_{cd} \ddot{x}^d. \tag{3.57}$$

Contracting with the inverse of the matrix g_{cd} allows us to rewrite the above as

$$\boxed{\ddot{x}^d + \Gamma^d_{ab} \dot{x}^a \dot{x}^b = 0} \tag{3.58}$$

where the field Γ^d_{ab} is defined by

$$\boxed{\Gamma^d_{ab} = \tfrac{1}{2} g^{dc} \left(\partial_a g_{cb} + \partial_b g_{ca} - \partial_c g_{ab} \right)}. \tag{3.59}$$

In physics, this field is called the Levi-Civita connection, from the Italian mathematician Tullio Levi-Civita, or the Christoffel symbols, from the German mathematician Elwin Bruno Christoffel (mathematicians use these denominations in a more nuanced manner). A geodesic, namely a shortest line between two points, satisfies the geodesic equation (3.58) in the parametrisation where (3.54) holds.

- **Exercise:**

Write the equations that describe a parallel and a meridian on a 2-sphere. Show explicitly whether or not they solve the geodesic equation. Compute their length using the equation of the length for a geodesic. Study the same problem for a 3-sphere.

3.2.2   Fields and Derivatives on Riemann Spaces

• Scalar fields

On a Riemann space, we can define scalar fields $\varphi(x)$ that associate a real number to each point of the space. If we use a different set of coordinates $x \to \tilde{x}(x)$, the same field is obviously expressed by a different function, related to the previous one by

$$\tilde{\varphi}(\tilde{x}) = \varphi(x(\tilde{x})). \tag{3.60}$$

• Vector fields

While the definition of scalar fields on a Riemann space is immediate, the definition of vector fields is more tricky. The reason is that a vector on a point p of a 2d surface Σ immersed in R^3 is an element of the space T_p tangent to Σ at p, a fact well known from elementary geometry. The tangent space is a plane in the R^3 space where Σ is embedded. But a Riemann space is defined independently of any embedding. So, what is a vector at a point of the space?

The definition of vector field common in physics textbooks is a bit twisted. Let me start with two examples.

First example. Consider the quantity $w_a(x)$ defined by the derivatives of a scalar field:

$$w_a(x) \equiv \frac{\partial \varphi(x)}{\partial x^a}. \tag{3.61}$$

It is easy to see, using the Leibniz rule for derivation *[do it!]*, that in different coordinates this reads

$$\tilde{w}_a(\tilde{x}) = \frac{\partial x^b}{\partial \tilde{x}^a} w_b(x(\tilde{x})). \tag{3.62}$$

Any field that transforms under change of coordinates as w_a is called a *covariant vector field* and is indicated with a downstairs index.

Second example. On the other hand, imagine we have a flow of particles moving in space and at each point x the velocity is $v^a(x) = \frac{dx^a}{dt}$. If we change coordinates, again the Leibniz rule gives *[do it!]*

$$\tilde{v}^a(\tilde{x}) = \frac{\partial \tilde{x}^a}{\partial x^b} v^b(c(\tilde{x})), \tag{3.63}$$

which is different from (3.62) (the new coordinates are upstairs in the Jacobian, not downstairs). Any field that transforms as v^a under change of coordinates is called a *controvariant vector field*, and it is indicated with an upstairs index.

It is easy to see, using the various transformation rules, that if $v^a(x)$ is a controvariant field, then $v_a(x) \equiv g_{ab}(x)v^b(x)$ is a covariant field. It is customary to indicate with the same letter the covariant and controvariant fields related in this manner. They have, in a sense, the same information, if the metric $g_{ab}(x)$ is known.

• **Tensor fields**

By definition, a 'tensor' is a field with some downstairs indices (that transform as in (3.62)) and some upstairs indices (that transform as in (3.66))

$$\tilde{T}^{a\cdots}{}_{b\dots}(\tilde{x}) = \frac{\partial \tilde{x}^a}{\partial x^c} \cdots \frac{\partial x^d}{\partial \tilde{x}^b} \cdots \tilde{T}^{c\cdots}{}_{d\dots}(x(\tilde{x})). \tag{3.64}$$

The metric g_{ab} is a tensor: (3.30) is a special case of (3.64).

An equation between tensors is called a tensorial equation. If a tensorial equation holds in one coordinate system, it holds in all coordinate systems, because since the transformation of a tensor is linear, a tensor that vanishes in one coordinate system vanishes in all coordinate systems.

Careful: not all fields with indices are tensors. For instance, the Christoffel symbol Γ^a_{bc} is not a tensor. The way it transforms under a change of coordinates can be derived from the definition *[do it!]* and is not given by (3.64).

A bit of cleaner maths. The definition of vector fields (and tensors) given above (quantities with indices that transform in a certain manner under change of coordinates) is customary in physics textbooks. I have always found it awkward. So, for once I prefer to mention a cleaner definition given by mathematicians.

Intuitively, the tangent space is the space of the 'directions with length' at the point. The mathematician's solution to make this notion precise is to define a (controvariant) vector v at a point p as a derivative operator acting on scalar fields $\varphi(x)$ at p, measuring how they change in a certain direction. Its general form in terms of the coordinates x^a is

$$v \equiv v^a \left. \frac{\partial}{\partial x^a} \right|_{x_p} , \tag{3.65}$$

where x_p are the coordinates of p. The set of these derivative operators at p form a vector space (with coordinates v^a), which is *by definition* the tangent space T_p at p. The quantities v^a are the components of the vector v in the basis of the vectors $e_{(a)} \equiv \left. \frac{\partial}{\partial x^a} \right|_{x_p}$. The components change under a change of coordinates as

$$\tilde{v}^a = \frac{\partial \tilde{x}^a}{\partial x^b} v^b , \tag{3.66}$$

as follows easily from the Leibniz rule *[do it!]*.

A 'covariant' vector w is an element of the space T_p^* dual to T_p, namely a linear map on T_p. Given coordinates x^a, the map is $w(v) = w_a v^a$. Thus a coordinate system x^a defines the components w_a of a covariant vector, written with downstairs indices.

It is interesting to notice that in this language the metric g is a map $g : T_p \to T_p^*$. That is, it is a preferred identification of the tangent space with its dual. In components, at each point $w = g(v)$ reads $w_a = g_{ab} v^b$. This defines a scalar product on the tangent space: $(v, v') \equiv g(v)(v')$, hence a norm $|v|^2 = (v, v)$; therefore, it defines the length of any curve γ, obtained by integrating (along the curve γ) the norm of the tangent $\dot{\gamma}$, which is defined by the derivative operator $\dot{\gamma}(\varphi) = d\varphi(\gamma(t))/dt$.

A Riemann space is therefore nothing else than a space (a manifold) with a canonical identification between each tangent space and its dual.

- ● *Covariant derivative*

Tensors are important because an equation between tensors true in one coordinate system is true in any other coordinate system. Hence it is a relation that does not depend on a particular coordinate system.

Importantly, in general the derivative of a tensor is not a tensor. For instance, the quantity $\partial_a v^b \equiv \frac{\partial v^a}{\partial x^b}$ is not a tensor. This is because the derivative of the transformed tensor includes also the derivative of the Jacobian $\frac{\partial \tilde{x}^a}{\partial x^b}$, which in general is not a constant. However, it turns out that the quantity

$$D_a v^b \equiv \partial_a v^b + \Gamma^b_{ac} v^c \tag{3.67}$$

happens to be a tensor, as an explicit calculation shows *[do it!]*. This quantity is called the 'covariant derivative' of v^a. The covariant derivative of a covariant tensor is defined with a minus sign:

$$D_a w_b \equiv \partial_a w_b - \Gamma^c_{ab} w_c. \tag{3.68}$$

And the covariant derivative of a tensor with more indices picks one such term for each index. That is, for instance,

$$D_a w^b{}_c \equiv \partial_a w^b{}_c + \Gamma^b_{ad} w^d{}_c - \Gamma^d_{ac} w^b{}_d. \tag{3.69}$$

The covariant derivative is a coordinate-independent notion. If a tensor field $T^b{}_a = D_a v^b$ is the covariant derivative of the vector field v^a in a coordinate system, it is so in any coordinate system. This is not true for a relation like $\partial_a v^b = T^b_a$.

If a vector has a covariant derivative null along a path γ, namely $\dot{\gamma}^a D_a v^b = 0$, the angle between the vector and the tangent to the path remains constant along the path. This follows directly from the fact that this equation does not depend on the coordinates and is true in locally Cartesian coordinates. We say that such a vector is 'parallel transported' along the path. This result clarifies the meaning of Γ^a_{bc}: it tells us how to parallel transport vectors.

Another result that follows immediately from the fact that the equation is coordinate-independent and true in locally Cartesian coordinates is

$$D_a g_{bc} = 0. \tag{3.70}$$

3.2.3 Riemann Curvature

Armed with these tools, let's ask the key question. When is the intrinsic geometry defined by a metric field $g_{ab}(X)$ flat? Certainly the constant metric field

$$g_{ab}(X) = \delta_{ab} \tag{3.71}$$

defines a flat geometry, because, if this is the case, the distances are given by

$$ds^2 = \delta_{ab} \, dX^a dX^b, \tag{3.72}$$

which means that the coordinates X^a are the Cartesian coordinates of a Euclidean space. But if we introduce new coordinates x^a, the resulting metric

$$g_{ab}(x) = \delta_{cd}\frac{\partial X^c}{\partial x^a}\frac{\partial X^d}{\partial x^b} \tag{3.73}$$

also is flat, because the intrinsic geometry does not change by changing coordinates. How do we know if a given $g_{ab}(x)$ defines a flat intrinsic geometry?

To answer this question, let us compute the commutator of two covariant derivatives acting on a vector field *[do it!]*. We obtain

$$D_a D_b v^c - D_b D_a v^c = R^c{}_{dab}v^d, \tag{3.74}$$

where

$$\boxed{R^a{}_{bcd} = \partial_c \Gamma^a_{bd} - \partial_d \Gamma^a_{bc} + \Gamma^a_{ce}\Gamma^e_{bd} - \Gamma^a_{de}\Gamma^e_{bc}}. \tag{3.75}$$

Since the left-hand side of (3.74) is a tensor, so is the right-hand side; hence $R^a{}_{bcd}$ is a tensor (in spite of the fact that Γ^a_{bc} is not a tensor: it transforms in a far more complicated manner).

Now if the space is flat, we can choose global Cartesian coordinates. In these coordinates, $g_{ab}(x) = \delta_{ab}$, the quantity Γ^c_{ab} vanishes (because it is just made of derivatives of the metric), hence $R^a{}_{bcd} = 0$. But $R^a{}_{bcd}$ is a tensor: if it vanishes in one coordinate system, it vanishes in all. Hence we have found a way of testing if the space is flat: it must satisfy $R^a{}_{bcd} = 0$.

Riemann was able to show that $R^a{}_{bcd} = 0$ is not only necessary but also sufficient for the space to be flat. In other words, it is possible to change coordinates and put the metric in the form (3.71) in a region if and only if $R^a{}_{bcd} = 0$ in this region. (In analytical terms, $R^a{}_{bcd} = 0$ are the integrability conditions for solving (3.73) for $X^a(x)$, if $g_{ab}(x)$ is given.)

$R^a{}_{bcd}$, defined in (3.75), is called Riemann curvature, or Riemann tensor, the generalisation of the Gauss curvature to the intrinsic geometry of spaces of arbitrary dimension. This is the beautiful result of Riemann's thesis.

- **Geometrical meaning and properties of the Riemann curvature**

The Riemann tensor is an important quantity. A way to think about it is to separate the first two indices from the second two. The second two indices (which are anti-symmetric, as is obvious from the definition) locally determine a plane (for instance, when these indices are $(c = 1, d = 2)$ they determine the plane of the coordinates x^1 and x^2). The first two indices then give an infinitesimal rotation matrix. The geometry is the following. Recall the definition (3.1) of curvature in two dimensions: it gives the rotation angle of a vector parallel transported around a loop, divided by the area (in the limit of small area). In higher dimensions, the second two indices specify the surface in which the loop lies, and the first two indices give the infinitesimal rotation the vector undergoes when parallel transported around this loop. We shall see below an equation that expresses this fact.

At a given *single* point x it is always possible to choose coordinates in which the metric has the form $g_{ab}(x) = \delta_{ab}$ at this point. It suffices to diagonalise $g_{ab}(x)$ and scale the coordinates appropriately.

In fact one can do even better at a given *single* point x: it is always possible to choose coordinates where the metric field has this Euclidean form *and* all its first derivatives vanish (hence Γ^a_{bc} vanishes). These are the local Cartesian coordinates at the point. We can do so because, to first order in the distance, we can always approximate the geometry with a tangent space and use Cartesian coordinates on this. Curvature is seen only in the *second* derivatives of the metric tensor.

- **Geodetic deviation**

Consider two nearby geodesics that are momentarily parallel, namely the derivative of their distance is zero. In flat space these are parallel straight lines and therefore never get closer. In a Riemannian spacetime, obviously this is not anymore true (think of a sphere). If we call δx^a their separation, $\dot{x}^a$ the tangent, and $Dv^b/D\tau = dv^b/d\tau + \dot{x}^a \Gamma^b_{ac} v^c$ the covariant derivative along the geodesic, then the following holds

$$\frac{D^2}{D\tau^2} \delta x^a = R^a{}_{bcd}\, \delta x^c \dot{x}^b \dot{x}^d. \tag{3.76}$$

This shows that the Riemann curvature captures precisely the convergence of geodesics in curved space.

- **Ricci curvature, Ricci scalar and Bianchi identities**

A theorem states that the Riemann curvature and its only non-vanishing contractions, which are

$$R_{ab} = R^c{}_{acb}, \quad R = g^{ab}R_{ab}, \tag{3.77}$$

are the only tensors that one can build out of the metric and its first and second derivatives. The first is called the Ricci tensor, and the second is called the Ricci scalar, from the Italian mathematician Gregorio Ricci-Curbastro. As we shall see, they play a role in general relativity.

Finally, an explicit calculation shows that the Riemann curvature satisfies the differential identities

$$D_e R^{ab}_{cd} + D_d R^{ab}_{ec} + D_c R^{ab}_{de} = 0. \tag{3.78}$$

These are called the Bianchi identities $\partial_e F_{cd} + \partial_d F_{ec} + \partial_c F_{de} = 0$ and are the analogue of the identities satisfied by the Maxwell tensor.

- **Differential forms**

Differential forms offer a convenient alternative mathematical language for general relativity. They simplify notation and provide a useful mathematical perspective. I introduce them here for completeness, even if I shall not use them much.

Covariant tensors $T_{abc...}$ with p indices that are fully anti-symmetric in all their indices are also called differential p-forms and indicated with the compact notation

$$T = \frac{1}{p!} T_{abc...} dx^a \wedge dx^b \wedge dx^c \ldots \tag{3.79}$$

where $dx^a \wedge dx^b \equiv dx^a dx^b - dx^b dx^a$. In d dimensions, $0 \leq p \leq d$, since there are no fully anti-symmetric tensors with more than d indices. The wedge product $V = T \wedge U$ between a p form and a q form is given by the tensor $V_{abc...def...} = \frac{(p+q)!}{p!q!} T_{[abc...} U_{def...]}$ anti-symmetrised in all the indices. For instance, if $T = T_a dx^a$ and $S = S_a dx^a$, their wedge product is

$$F \equiv T \wedge S = (T_a S_b - T_b S_A) dx^a \wedge dx^b. \tag{3.80}$$

The differential operator d, defined by the action on a p-form

$$(dT) = (p+1)\, \partial_a T_{bcd...} dx^a \wedge dx^b \wedge dx^c \wedge dx^d \ldots, \tag{3.81}$$

sends p-forms into $(p + 1)$-forms and satisfies the fundamental equation

$$d^2 = 0 \tag{3.82}$$

[show it!]. Scalar functions are zero-forms and covariant vectors are 1-forms. In three dimensions, 2-forms are polar vectors and the gradient, curl, and divergence operators are the d operators acting respectively on 0-,1-, and 2-forms [show it!]. Therefore, this language brings order into the funny structure of vector calculus.

The integral of a p-form T over a p-dimensional surface Σ_p

$$I = \int_{\Sigma_p} T \equiv \int_{\Sigma_p} T_{abc...} \, dx^a dx^b dx^c \ldots \tag{3.83}$$

does not depend on the coordinates. Hence it is a quantity well-defined geometrically. An important theorem that generalises several similar results of vector calculus in 3d is the Stokes theorem, due to the French mathematician Élie Joseph Cartan, that states that

$$\int_{\Sigma} d\omega = \int_{\partial\Sigma} \omega, \tag{3.84}$$

where ω is a p-form, Σ a $(p + 1)$ surface, and $\partial\Sigma$ its boundary.

The tetrad field e_a^i defines the 1-form

$$e^i = e_a^i \, dx^a. \tag{3.85}$$

1-forms are dual to vectors in the sense that they contract with vectors to give a number: $T(v)(x) \equiv T_a(x)v^a(x)$. The geodesic equation in form notation reads

$$\frac{d}{d\tau} e^i(\dot{x}) + \omega^i{}_j(\dot{x})e^j(\dot{x}) = 0, \tag{3.86}$$

where ω^{ij} is a 1-form called the spin connection, defined by the equation

$$de^i + \omega^i{}_j \wedge e^j = 0. \tag{3.87}$$

The spin connection defines a covariant derivative on tensors with internal indices

$$D_a v^i \equiv \partial_a v^i + \omega^i_{aj} v^j, \tag{3.88}$$

and a straightforward calculation shows that this quantity transforms as a tensor.

The Riemann tensor in form notation is the curvature of the spin connection, defined by

$$R^{ij} = d\omega^{ij} + \omega^i{}_k \wedge \omega^{kj} \tag{3.89}$$

and is related to the conventional Riemann tensor by

$$e_{bj}R^{ij}_{cd} = e^i_a R^a{}_{bcd}.$$ (3.90)

The contraction $R^i_a \equiv e^b_j R^{ij}_{ab}$ is therefore related to the Ricci tensor by $R^i_a = \delta^{ij} e^b_j R_{ab}.$

• **Cartan geometry**

The mathematician Élie Joseph Cartan defined an extension of Riemannian geometry, today called Cartan geometry, by two *independent* fields e^i and ω^{ij}. Since they are independent, (3.87) does not hold. Instead, we can define a quantity called torsion by

$$T^i = de^i + \omega^i{}_j \wedge e^j.$$ (3.91)

The two equations (3.91) and (3.89) that define torsion and curvature are called the first and second Cartan structure equations. A Cartan geometry with vanishing torsion is a Riemannian geometry.

The extension of a Riemannian geometry to a Cartan geometry is often considered when studying the coupling of the gravitational field to fermions, but so far it has played no direct role in physical applications.

• **Computing curvature and geodesics with forms**

As a simple example of computing with forms, let us compute the Ricci tensor of the unit sphere. The diad of a sphere can be taken to be

$$e^1 = d\theta, \qquad e^2 = \sin\theta\, d\phi,$$ (3.92)

which yields (3.34). Writing equation (3.87) explicitly gives

$$de^1 + \omega^{12} \wedge e^2 = 0,$$ (3.93)

$$de^2 + \omega^{21} \wedge e^1 = 0,$$ (3.94)

that is, using the value of the diad explicitly,

$$\omega^{12} \wedge \sin\theta\, d\phi = 0,$$ (3.95)

$$\cos\theta\, d\theta \wedge d\phi + \omega^{21} \wedge d\theta = 0.$$ (3.96)

The first equation gives $\omega^{12}_\theta = 0$, the second, $\omega^{12}_\phi = -\cos\theta$. That is,

$$\omega^{12} = -\cos\theta\, d\phi.$$ (3.97)

From the definition of the curvature, we have

$$R^{12} = d\omega^{12} + \omega^{1k} \wedge \omega^{k2} = \sin\theta\, d\theta \wedge d\phi,$$ (3.98)

while the second term vanishes. We easily see that the diagonal components of R^{ij} vanish as well. Finally, the Ricci scalar is

$$R = e_i^a e_j^b R_{ab}^{ij} = e_1^1 e_2^2 R_{12}^{12} + e_2^2 e_1^1 R_{21}^{21} = 2\frac{\sin\theta}{\sin\theta} = 2. \tag{3.99}$$

As a second exercise, let us see if the meridians or the parallels are geodesic. The meridians are given by $x^a = (s, \phi_0)$, thus $\dot{x}^a = (1, 0)$. Hence $e^i(\dot{x}) = (1, 0)$ and $\omega^{12}(\dot{x}) = 0$, which clearly solves the geodesic equation. The parallels are given by $x^a = (\theta_0, s/\sin\theta_0)$, where the factor $1/\sin\theta_0$ ensures that ds is the proper length. Thus $\dot{x}^a = (0, 1/\sin\theta_0)$. Hence $e^i(\dot{x}) = (0, 1)$ and

$$\omega^{12}(\dot{x}) = -\cos\theta_0/\sin\theta_0, \tag{3.100}$$

so that the geodesic equation reads

$$e^1(\dot{x}) + \omega^{12}(\dot{x}) \wedge e^2(\dot{x}) = 0 \tag{3.101}$$
$$e^2(\dot{x}) + \omega^{21}(\dot{x})e^1(\dot{x}) = 0. \tag{3.102}$$

The first equation vanishes identically. The second has only the second term, which gives

$$-\cos\theta_0 d\phi \wedge \sin\theta_0 d\theta = 0, \tag{3.103}$$

which vanishes only when $\cos\theta_0 \sin\theta_0$ vanishes, namely on $\theta_0 = 0, \pi/2, \pi$. This means that (correctly) the only parallels that are geodesics are the equator and the two degenerate parallels at the North and South Poles.

3.3 GEOMETRY

Consider a 3d space with triad field $e_a^i(x)$ and metric $g_{ab}(x)$. Geometrical quantities can be expressed as follows.

As we have already seen, the length of a line γ defined by $x^a(\tau)$ is

$$L[\gamma] = \int_\gamma \sqrt{g_{ab}\frac{dx^a}{d\tau}\frac{dx^b}{d\tau}}\,d\tau = \int_\gamma \sqrt{e^i(\dot{x})e^j(\dot{x})\delta_{ij}}\,d\tau. \tag{3.104}$$

The volume of a finite region R is given by

$$V[R] = \int_R \sqrt{\det[g]}\,d^3x = \int_R \det[e]\,d^3x = \frac{1}{3!}\int \epsilon_{ijk}\,e^i \wedge e^j \wedge e^k. \tag{3.105}$$

Consider next a 2d surface S immersed in the space coordinated by the coordinates σ, τ and defined by the functions $x^a(\sigma, \tau)$. The tangents to the surface are

$$\dot{x}^a = \frac{\partial x^a}{\partial \sigma}, \quad \hat{x}^a = \frac{\partial x^a}{\partial \tau}. \tag{3.106}$$

The area of the surface is given by the integral of the square root of the determinant $g^{(2)}$ of the metric induced on the surface

$$
\begin{aligned}
A[S] &= \int_S \sqrt{\det g^{(2)}} \, d\sigma \, d\tau \\
&= \int_S \sqrt{(g_{ac}g_{bd} - g_{ab}g_{cd})\dot{x}^a \hat{x}^b \dot{x}^c \hat{x}^d} \, d\sigma \, d\tau \\
&= \int_S \sqrt{E^i_{ab}E^i_{cd} \dot{x}^a \hat{x}^b \dot{x}^c \hat{x}^d} \, d\sigma \, d\tau \\
&\equiv \int_S |E|,
\end{aligned}
\tag{3.107}
$$

where the two-form

$$E^i = E^i_{ab} dx^a \wedge dx^b = \frac{1}{2}\epsilon_{ijk} e^j \wedge e^k \tag{3.108}$$

is called the 'Ashtekar's electric field', or the 'gravitational electric field'. Hence we can say that 'the area is the norm of the gravitational electric field'.

3.3.1   Lorentzian Geometry

Riemannian geometry has the property that it is well approximated by a *Euclidean* space at each point. This is the mathematics that Einstein found ready to use for describing curved spaces. But it was not exactly what Einstein needed. What he needed was a small variant of the Riemannian geometry: a mathematics that could describe curved *spacetimes*; that is, spaces well approximated at each point by *Minkowski* space.

This is easy to do. It suffices to replace (3.18) with

$$g_{ab}(x) = \eta_{ij} \, e^i_a(x)e^j_b(x), \tag{3.109}$$

where $\eta_{ij} = \text{diag}[-1, 1, 1, 1]$ is the Minkowski metric.

The spaces defined by the metric (3.109) are called pseudo Riemannian or Lorentzian. The difference with a Riemannian space is simply the fact that the signature of the metric is $(-, +, +, +)$ instead of being $(+, +, +, +)$; namely, at each point the matrix $g_{ab}(x)$ has one negative and three positive eigenvalues.

In dealing with Lorentzian spaces, we should take care to distinguish upper and lower internal indices $i, j, \ldots$, which are raised and lowered with the Minkowski metric as we do in special relativity because of the minus sign of η_{00}.

In a Lorentzian geometry, intervals can be timelike, spacelike, or null, according to whether (3.17) is negative, positive, or zero. A line is timelike, spacelike, or null if $|\dot{x}|^2$ is everywhere, respectively, negative, positive, or null.

Lorentzian spaces have therefore an interesting structure of local light cones: there is a light cone at every point, and these light cones vary smoothly from one point to the other. See Figure 3.7.

If there are no closed timelike curves, this structure defines a partial ordering among its points, as in Minkowski space. That is, each point has a 'future' and a 'past' (the two regions that can be reached with timelike lines), and a region which is neither past nor future (the set of points at spacelike separation from the given point, which can be called the 'extended present').

FIGURE 3.7 Lorentzian geometry: there is a local light cone at each point. The tangent of a timelike (dotted), null (dashed), or spacelike (continuous) line is, respectively, inside, on, or outside these cones.

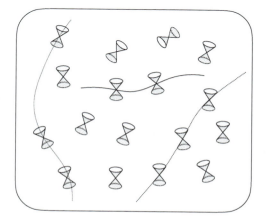

We will see genuinely curved Lorentzian spaces in the following chapters. Below I list some examples of general coordinates on Minkowski space: they will all be useful for understanding general relativistic physics.

- **Polar spatial coordinates**

A trivial example of a change of variable in a Lorentzian metric is to write the Minkowski metric in polar spatial coordinates (t, r, θ, ϕ). This gives

$$ds^2 = -dt^2 + dr^2 + r^2 d\theta^2 + r^2 \sin^2\theta d\phi^2 = -dt^2 + dr^2 + r^2 d\Omega^2.$$
(3.110)

- **Rindler coordinates**

On a two-dimensional spacetime with Minkowski coordinates t, x, and metric $ds^2 = -dt^2 + dx^2$, there is an important spacetime analogue of polar coordinates: the coordinates $\tau \in [-\infty, \infty]$, $\rho \in [0, \infty]$ defined by

$$t = \rho \sinh\tau, \qquad x = \rho \cosh\tau.$$
(3.111)

These are called Rindler coordinates. They are like polar coordinates but adapted to the Minkowski signature. $\rho = \sqrt{x^2 - t^2}$ is the invariant 4-distance from the origin, and the constant τ lines are the instantaneous simultaneity surfaces of observers that move at constant acceleration along the constant ρ lines. The lines of equal ρ and equal τ are drawn in Figure 3.8. In these coordinates, the metric reads *[compute it]*

$$ds^2 = d\rho^2 - \rho^2 d\tau^2.$$
(3.112)

Unlike the usual polar coordinates, however, these coordinates cover only a portion of Minkowski space: the 'Rindler wedge' $x > |t|$. See Figure 3.8. Notice that a timelike line (such as $r = 1$) crosses the light ray $x = r$ only when τ goes to infinity. This metric therefore does not cover the entire spacetime. The portion covered is bounded by two distinct lines: the outgoing null light ray $x = t > 0$ and the ingoing null light ray $x = -t > 0$. Recalling this will be very useful when studying black holes.

- **Null coordinates**

In Lorentzian spaces it is sometimes useful to use null coordinates. These are coordinates that track light rays. For instance, in a 2d Minkowski spacetime with metric $ds^2 = -dt^2 + dx^2$, we can introduce the null coordinates

$$U = t - x, \qquad V = t + x.$$
(3.113)

Clearly $U = constant$ and $V = constant$ are null lines. Light rays follow these lines. Differentiating these and inserting the line element, we have

$$ds^2 = -dU\, dV.$$
(3.114)

FIGURE 3.8 Rindler coordinates on the
Rindler wedge

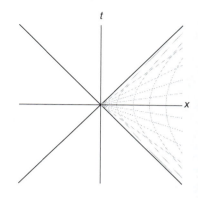

In four dimensions, the Minkowski metric can be written in the form

$$ds^2 = -dU\,dV + r^2 d\Omega^2,$$ (3.115)

where $r = r(U, V) = (V - U)/2$.

Null coordinates that cover only the Rindler wedge are given by

$$u = \log U, \quad v = -\log(-V).$$ (3.116)

In these coordinates, the metric reads

$$ds^2 = -e^{v-u}du\,dv,$$ (3.117)

and the two light rays $t = x$ and $t = -x$ are pushed to $u \to -\infty$ and $v \to \infty$. See Figure 3.9.

These different sets of coordinates on Minkowski space will help us understand what will happen with black holes.

Sometimes one also uses mixed coordinates such as $v = (t + x)/2, r$, or, alternatively, $u = (t - x)/2, r$, and the metric reads

$$ds^2 = -dv^2 - 2dv\,dr + r^2 d\Omega^2$$ (3.118)

and

$$ds^2 = -du^2 + 2du\,dr + r^2 d\Omega^2.$$ (3.119)

Coordinates analogous to these will also play a role in understanding the geometry of black holes.

- **Rotating coordinates**

Consider a platform rotating slowly with angular velocity ω. Let t, r, ϕ be the time and spatial polar coordinates. The metric in these coordinates is

$$ds^2 = -dt^2 + dr^2 + r^2 d\phi^2.$$ (3.120)

If φ is an angular coordinate on the rotating platform itself, we have

$$\phi = \varphi - \omega t.$$ (3.121)

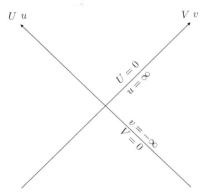

FIGURE 3.9 The null coordinates U, V cover the full Minkowski space. The null coordinates u, v cover only the Rindler wedge (in grey).

Therefore, the metric in coordinates attached to the platform is

$$ds^2 = -(1 - r^2\omega^2)dt^2 + dr^2 + r^2 d\varphi^2 - 2r^2\omega\, d\varphi dt. \qquad (3.122)$$

The non-diagonal term $-2r^2\omega\, d\varphi\, dt$ characterises the rotating frame: it indicates that this frame rotates with respect to an inertial frame. (The correction term $1 - r^2\omega^2$ is a relativistic time dilatation due to the motion with respect to the inertial frame.)

We are now armed with all the maths we need. We can finally get to physics.

PART II The Theory

4   Basic Equations

4.1   GRAVITATIONAL FIELD

The gravitational field is described by a tetrad $e_a^i(x)$ (to be compared with the Maxwell field $A_a(x)$) or equivalently by the 4d Lorentzian metric

$$g_{ab}(x) = \eta_{ij}\, e_a^i(x) e_b^j(x). \tag{4.1}$$

The geometry of spacetime is determined by the line element

$$ds^2 = g_{ab}(x)\, dx^a dx^b. \tag{4.2}$$

At each spacetime point p with coordinates x_p^a, the tetrad determines local Cartesian coordinates $X^i = e_a^i(x)(x^a - x_p^a)$ that define a 'free falling' local reference frame, where physics is just as it is in Minkowski space up to the second order in X^i. The existence of this reference system, which put Einstein on the right track to find the theory, is called the 'equivalence principle'.

Since coordinates are arbitrary, it makes little sense to attribute physical dimensions to them. Hence we see from the last equation that the gravitational field $g_{ab}(x)$ has the natural dimension of $Length^2$ (we have $c = 1$). A different convention can also be found in the literature: to have $g_{ab}(x)$ dimensionless and to insist on having dimension-full coordinates. This comes naturally when $g_{ab}(x)$ is the Minkowski metric. Far less so, say, using polar coordinates.

From the equivalence principle, it follows that a clock moving along a finite timelike line $\gamma : \tau \mapsto x^a(\tau)$ measures the time

$$T = \int_\gamma \sqrt{-g_{ab}\dot{x}^a\dot{x}^b}\; d\tau. \tag{4.3}$$

In a region small with respect to the curvature scale, T is the Lorentz time in the frame of the clock *[show it!]*. It is proportional to the number of oscillations of any small harmonic oscillator (= a 'clock'). T is called the 'proper time' along the line, to be distinguished from the 'coordinate time' t.

The length of a rod sitting on a finite spacelike line γ is

$$L = \int_{\gamma} \sqrt{g_{ab}\dot{x}^a \dot{x}^b} \; d\tau. \tag{4.4}$$

This quantity is proportional to the number of atoms in a solid structure on scales smaller than the scale of the curvature (= a 'rod'). It is called 'proper length'.

Important. The relation of rods and clocks with $g_{ab}(x)$ does *not* need to be postulated, as sometimes stated: it follows from the dynamics of these physical devices, coupled to the gravitational field. In fact, rods and clocks are just those devices that couple to gravity measuring ds. This point is important conceptually: geometry is not a Kantian a priori, necessary to conceive the world: *geometry is an epiphenomenon of the gravitational field*.

4.2   EFFECTS OF GRAVITY

The following derives immediately from the equivalence principle. Massive particles (not subjected to forces other than gravity) move along geodesics. That is, in the parametrisation of their world line $x^a(\tau)$ where $|\dot{x}|^2 = -1$, the equation of motion of a massive particle is the geodesic equation

$$\ddot{x}^d + \Gamma^d_{ab}\dot{x}^a \dot{x}^b = 0 \tag{4.5}$$

(to be compared with the Lorentz force equation $\ddot{x}^d - \frac{e}{m}F^d{}_a\dot{x}^a = 0$) where the Levi-Civita connection is defined in (3.58), which I copy here for completeness:

$$\Gamma^d_{ab} = \frac{1}{2}g^{dc}\left(\partial_a g_{cb} + \partial_b g_{ca} - \partial_c g_{ab}\right). \tag{4.6}$$

The second term in the geodesic equation (4.5) can be viewed as the 'gravitational force term', or as the effect of the spacetime geometry on the motion. The two notions are identified. A particle moving along a geodesic is called in 'free fall'. If other forces act on the particle, they add to (4.5). For instance, if the particle is charged,

$$\ddot{x}^d + \Gamma^d_{ab}\dot{x}^a\dot{x}^b - \frac{e}{m}F^d{}_a\dot{x}^a = 0, \tag{4.7}$$

where F_{ab} is the electromagnetic field.

The trajectory of light rays is simply determined by $ds = 0$. That is, light rays (electromagnetic wavefront trajectories in the high-frequency limit) move along null lines: their world lines $x^a(\tau)$ satisfy

$$\frac{ds}{d\tau} = |\dot{x}| = \sqrt{g_{ab}(x)\,\dot{x}^a\dot{x}^b} = 0. \tag{4.8}$$

The effect of gravity on any other system (beside particles) can be obtained by replacing η_{ab} with $g_{ab}(x)$ and the derivatives ∂_a with covariant derivatives D_a in the equations of motion of the system. For instance, the interaction between gravity and electromagnetism is given by the Maxwell equations coupled to gravity

$$D_a F^{ab} = \partial_a F^{ab} + \Gamma^a_{ad}F^{db} + \Gamma^b_{ad}F^{ad} = 4\pi J^b. \tag{4.9}$$

The interaction with a Dirac field cannot be written in terms of the metric; it requires the tetrad. For this reason the tetrad formalism, although historically later, is more fundamental. The Dirac equation in the presence of gravity becomes

$$\gamma^i e^a_i \partial_a \psi + \omega^{ij}_a \gamma_i \gamma_j \psi = 0, \tag{4.10}$$

where γ_i are the Dirac matrices. The notation can be made compact as

$$D\!\!\!/\,\psi = 0. \tag{4.11}$$

4.3   FIELD EQUATIONS

The field equations (to be compared with the Maxwell equations $\partial_a F^{ab} = 4\pi J^b$) are the Einstein equations:[1]

$$\boxed{R_{ab} - \tfrac{1}{2}Rg_{ab} + \lambda g_{ab} = 8\pi G T_{ab}}. \tag{4.12}$$

---

[1] Einstein, Albert (1915), 'Die Feldgleichungen der Gravitation', Sitzungsberichte der Preussischen Akademie der Wissenschaften zu Berlin: 844–847 (without λ). Einstein, Albert (1917), 'Kosmologische Betrachtungen zur allgemeinen Relativitätstheorie', Sitzungsberichte der Preussischen Akademie der Wissenschaften: 142 (with λ).

Many books have 'derivations' of these equations. These may illuminate aspects of them, such as the fact that they are singled out by certain sets of assumptions. But if these equations could have been simply 'derived': Einstein would not have taken years – and a remarkable sequence of articles with *wrong* equations – before finding them.

In terms of tetrads, the Einstein equations read

$$R_a^i - \frac{1}{2} R e_a^i + \lambda e_a^i = 8\pi G T_a^i. \tag{4.13}$$

Two constants enter these equations. G is the Newton constant, and λ is the cosmological constant. Their currently measured values are

$$G \sim 6.67 \times 10^{-8} \, \frac{\text{cm}^3}{\text{g s}^2}, \qquad \lambda \sim 1.11 \times 10^{-48} \, \frac{1}{\text{cm}^2}. \tag{4.14}$$

The first has been roughly known since Newton, but it is today one of the least precisely measured fundamental quantities in physics. The second has been measured only in the last decade. Before this time, it was erroneously expected to be zero or negative by many theoreticians (not all of them).

Some physicists, especially those trained in particle physics, think that the bare value of λ should be zero and its measured value is a purely quantum effect due to the radiative corrections. There is no reason to expect so. Because of this mistake, many are confused by this constant and claim, mistakenly, that it represents a dark mystery, called 'dark energy'. The constant λ is no more or less mysterious than others of the twenty or so fundamental constants in current fundamental physics. (For instance, in perturbation theory around flat space, the quantum field theoretical radiative corrections to λ diverge polynomially in the cutoff, just as for the mass of the Higgs boson.)

4.4 SOURCE IN THE FIELD EQUATION

The source sitting on the right-hand side of the Einstein equations, T_{ab}, is the energy momentum tensor of matter. This is a field that describes the density and flow of energy and momentum in spacetime. It is the gravitational analogue of the electromagnetic current. The time-time component of the tensor with upper indices, T^{oo}, gives the energy density, the time-space components give the energy flow and the momentum density, and the space-space components give the flow of momentum. Let's see in more detail how this field is defined.

For a single particle moving along a world line γ, it is

$$T^{ab}(x) = m \int_\gamma d\tau \, \dot{x}^a \dot{x}^b \delta(x, x(\tau)), \qquad (4.15)$$

where $\delta(x, y)$ is Dirac's delta distribution, defined by

$$\int d^4x \, f(x) \, \delta(x, y) = f(y). \qquad (4.16)$$

That is, $T^{ab}(x)$ is concentrated along the particle's world line, and it is proportional to $m\dot{x}^a \dot{x}^b = p^a \dot{x}^b$. In the rest frame of the particle $\dot{x}^a = (1, 0, 0, 0)$, therefore, there is no momentum and no energy flow and the energy momentum tensor has the only component $T^{oo}(x) \sim m$ concentrated on the particle's position.

The energy momentum tensor of electromagnetism is

$$T^{ab} = F^{ac} F^b{}_c - \frac{1}{4} g^{ab} F^{cd} F_{cd}, \qquad (4.17)$$

where, recall, indices are raised and lowered with the metric tensor. The general recipe for computing T^{ab} and the source of these definitions is in the next chapter.

4.5   VACUUM EQUATIONS

In general relativity a region where $T_{ab} = 0$, namely without matter, is called 'vacuum'. Here, neglecting the cosmological constant, the Einstein equations read

$$R_{ab} - \frac{1}{2} R g_{ab} = 0. \qquad (4.18)$$

Contracting with g^{ab}, we have $R = 0$. Substituting this back into (4.18) gives

$$R_{ab} = 0. \qquad (4.19)$$

That is, in the absence of matter the Ricci tensor vanishes. This is not a sufficient condition for flatness. Hence, in general, spacetime is curved also in the absence of matter. Matter is not sufficient to determine the gravitational field, just as charges are not sufficient to determine the electromagnetic field. A spacetime where the Ricci tensor is everywhere vanishing is called an Einstein space.

So: (Riemann = 0)↔ flat space. (Ricci = 0)↔ empty space.

These equations define general relativity. They are sufficient to predict gravitational waves, black holes, the expansion of the universe, the Big Bang, to ground the GPS technology, and all that.

5 Action

• **The Einstein–Hilbert action**

All the dynamical equations of the previous chapter follow from an action principle. In a matter of weeks or perhaps even days after Einstein's completion of the field equations, David Hilbert, who was competing with Einstein for completing the theory, found an action from which the Einstein equations can be derived. It is very simple (here in the form including λ):

$$S[g] = \frac{1}{16\pi G} \int \sqrt{-g}\,(R - 2\lambda), \tag{5.1}$$

where $g = \det(g_{ab})$. The derivation of the field equations from the variation of this action is reported in many books, and I do not repeat it here.

The same field equations can also be derived in form notation, considering the triad field e and the spin connection ω as independent variables, from the action

$$S[e, \omega] = \int \epsilon_{ijkl}\, R^{ij} \wedge e^{k} \wedge e^{l}, \tag{5.2}$$

where R^{ik} is the curvature of ω, defined in (3.88). Here I have dropped the cosmological constant term and overall multiplicative constant for simplicity. Varying ω in this action gives equation (3.87), namely the vanishing of the torsion (the equation that states that ω is the spin connection defined by e). Varying e gives the Einstein equations.

It is possible to add another term to the action without changing the field equations:

$$S[e, \omega] = \int \epsilon_{ijkl}\, R^{ij} \wedge e^{k} \wedge e^{l} + \frac{1}{\gamma} R_{ij} \wedge e^{i} \wedge e^{j}. \tag{5.3}$$

This extra term does not modify the classical theory but has an effect on quantum theory in loop quantum gravity, as we shall see. The dimensionless coupling constant γ is called the 'Barbero–Immirzi parameter', or 'Immirzi parameter', and it is commonly (but not always) assumed to be of the order unit.

- *Matter actions*

The action of a particle moving in a given gravitational field is extremely simple. Since a particle follows a geodesic, and a geodesic is an extremum of the length, the action is simply proportional to the 4d length of the trajectory, namely the proper time along it. The proportionality constant is the mass, which gives the correct coupling with the field:

$$S = m \int ds. \tag{5.4}$$

That is, explicitly,

$$S[x] = m \int \sqrt{g_{ab}(x(\tau)) \frac{dx^a(\tau)}{d\tau} \frac{dx^b(\tau)}{d\tau}} \, d\tau. \tag{5.5}$$

Notice that the physical trajectory is a *maximum* of the action (as is clear in the Minkowski case): the inertial trajectory is the trajectory where a clock measures the *longest* time between two spacetime points.

The equivalence principle implies that the action of any system interacting with gravity can be obtained from the action of the system in special relativity by replacing partial derivatives with covariant derivatives and the Minkowski metric with the gravitational field $g_{ab}(x)$, together with the replacement of the volume element d^4X with the invariant volume element $\sqrt{-g}\, d^4x$, where g is the determinant of g_{ab}.

The quantity $\sqrt{-g}\, d^4x$ is called the invariant volume because it does not change when changing coordinates: the coordinate volume element d^4x and the determinant of the metric transform as opposite powers of the Jacobian of the coordinate transformation [show it!] .

For instance, the action of the electromagnetic field reads

$$S = \frac{1}{4} \int d^4x \sqrt{-g}\, F_{ab} F_{cd}\, g^{ac} g^{bd} \tag{5.6}$$

and the action of a Dirac field

$$S = \frac{1}{4} \int d^4x \sqrt{e}\, \bar{\psi} \slashed{D} \psi, \tag{5.7}$$

where e is the determinant of e^i_a.

The energy momentum tensor $T_{ab}(x)$ of a matter component can be computed, in general, as the variation of the action with respect to the metric. If the matter action is $S[\varphi, g_{ab}]$, the energy momentum tensor can be computed as

$$T^{ab}(x) = -\frac{2}{\sqrt{-g}}\frac{\delta S}{\delta g_{ab}(x)}.$$ (5.8)

This is why it appears in the right-hand side of the Einstein equations.

- **Exercise:**

Derive the dynamical equations of the previous chapter from the action given in this chapter.

6   Symmetries and Interpretation

Before starting to use the above equations to describe nature, let us pause to discuss their interpretation. The interpretation of the equations of general relativity is not anymore controversial, but it is subtle and has taken time to be clarified. For a long time, the best relativists (including Einstein) were confused about things like whether the gravitational waves are real or gauge artefacts (Einstein changed his mind twice about this), whether the surface of a black hole is the end of the world (Einstein got this wrong), whether the equations admit solutions without matter (Einstein was long wrongly convinced that they didn't), and similar. Let us, therefore, be careful.

- ### The 'meaning of the coordinates'
As Einstein himself put it, the difficulty is the 'meaning of the coordinates'. In physics before general relativity, coordinates localised points and events by giving their distance from fixed reference objects. For example, before general relativity, if I say that an object is at the position with Cartesian coordinates $X = 3, Y = 0, Z = 0$, I am saying that the object is at a distance $d = 3$ from the origin in the units I am using. In general relativity, coordinates do not have this meaning. If I say that an object is at the position with general coordinates $x = 3, y = 0, z = 0$, the object can be at any distance from the origin of the coordinates. Distance is determined by $g_{ab}(x)$, not by the coordinates. Since I can label events with *arbitrary* coordinates, to say, for instance, that an event is at the coordinates $x = 3, y = 0, z = 0, t = 5$ by itself carries no information at all.

- ### Symmetry
From a mathematical perspective, the freedom to choose coordinates is reflected in an under-determinacy of the evolution in the field

equations. If $g_{ab}(x)$ is a solution of Einstein's equations, then so is $\tilde{g}_{ab}(x)$ defined in (3.30), which I repeat here for completeness:

$$g_{ab}(x) \rightarrow \tilde{g}_{cd}(\tilde{x}) = \frac{\partial x^a}{\partial \tilde{x}^c} \frac{\partial x^b}{\partial \tilde{x}^d} g_{ab}(x(\tilde{x})), \qquad (6.1)$$

for any smooth invertible function $x^a(\tilde{x})$. This is a symmetry of the theory. It is called general covariance, or diffeomorphism invariance, depending on the way we look at it (as is discussed in the following pages).

- **Gauge**

If the change of coordinates $x^a \rightarrow \tilde{x}^a(x)$ is the identity before a certain time, we see immediately that the same gravitational field can evolve into two different metrics in the future (g_{ab} and $\tilde{g}_{ab}$), respecting the Einstein equations in both cases. This does not mean that the theory is indeterministic. It only means that the symmetry (6.1) must be interpreted as a gauge invariance: that is, solutions of the Einstein equations related by (6.1) describe the *same* physical spacetime. In the applications we shall see explicitly how different metrics actually describe the same physics.

Hence a physical spacetime is not described by a given field $g_{ab}(x)$ but rather by the equivalence class of these fields under the gauge transformation (6.1).

In the presence of matter, the transformation defining the equivalence class must also include the matter fields. For instance, if matter is described by the electromagnetic tensor F_{ab}, then physical histories of the world correspond to equivalence classes of solutions (g_{ab}, F_{ab}) of the coupled Einstein and Maxwell field equations, under a common coordinate transformation on g_{ab} and F_{ab}.

- **Background independence**

Diffeomorphism invariance reflects the fact that in general relativity there is no fixed background spacetime with respect to which events are located.

This is because the transformation (6.1) can be interpreted in two different manners:

1 **General Covariance.** The map $x^a \mapsto \tilde{x}^a(x)$ can be interpreted as a change of coordinates, namely as a relabelling of the points: the point with coordinates x^a is labelled with new coordinates $\tilde{x}^a$. Interpreted in this manner, the symmetry (6.1) is called general covariance.

2 **Diffeomorphism Invariance.** Alternatively, the map $x^a \mapsto \tilde{x}^a(x)$ can be interpreted as defining a map M from the manifold to itself. In this case, coordinates do not change, but the point p with coordinates x^a is sent by the map to the different point $\tilde{p} = M(p)$ with coordinates $\tilde{x}^a(x)$. Seen in this way, coordinates are irrelevant. In a coordinate-independent language, a Riemannian metric g assigns a distance $d_g(p, q)$ to any two points. The transformed metric $\tilde{g}$ is defined by the distance

$$d_{\tilde{g}}(p, q) \equiv d_g(M^{-1}(p), M^{-1}(q)). \tag{6.2}$$

Notice that the distance between the two points p and q is different in the two cases:

$$d_{\tilde{g}}(p, q) \neq d_g(p, q), \tag{6.3}$$

and yet the two metrics are physically indistinguishable. This seems mysterious at first and has nothing to do with coordinates.

The solution of this apparent puzzle is of major importance. The solution is that physical points are not defined by themselves. They are *only* defined by the solutions of the equations of motion, by the fields, by the positions of the particles, and by the geometry.

That is, location is defined only with respect to the dynamical physical fields (including the metric). This is profoundly different from what happens in non-general relativistic physics, where we assume that physical spacetime points are well defined (say by their distances from reference frame axes) independently from the dynamical fields.

For instance, imagine a compact space with a geometry that is – say – nearly spherical, except for a bump (a mountain) over the point P. Imagine then a compact space with a geometry that is nearly spherical, except for a bump (a mountain) over a different point Q. These two geometries represent *the same* physical configuration, because the location of the bump with respect to the background manifold has no physical meaning. Only relative location of dynamical entities has physical meaning. See Figure 6.1.

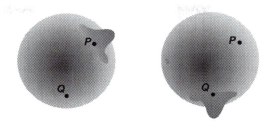

FIGURE 6.1 A simple example of background independence: two metrics, with a bump in two different locations, are physically indistinguishable, because in general relativity location is defined only by the geometry itself (and by any other dynamical field, if present). Hence the point P of the left panel must be identified with the point Q of the right: both are defined by the location of the bump.

6.1   TIME AND ENERGY

- **Different notions of time**

General relativity is formulated in terms of fields that depend on coordinates x^a which can include a 'time' coordinate $x^0 = t$. This formulation seems the same as that of non-general-relativistic field theories, such as Maxwell theory, but the similarity is misleading. While the t coordinate in Maxwell equations indicates the quantity measured by a clock, the coordinate t in Einstein's equations *does not indicate time measured by a clock*. In general, it does not indicate anything physical or measurable. The time measured by a clock moving along a curve $\gamma : \tau \to \gamma^a(\tau)$ between two events is, recall, the proper time

$$T = \int_\gamma \sqrt{-g_{ab}\frac{d\gamma^a}{d\tau}\frac{d\gamma^b}{d\tau}}d\tau. \qquad (6.4)$$

Since the proper time between two spacetime events depends on γ, in general there is no way to assign a unique physical time coordinate to spacetime events.

General relativity therefore treats time evolution in a profoundly different way from pre-general-relativistic physics. While

FIGURE 6.2 Which of the two clocks measures true time? Which of the two measures a longer time?

this can be seen as the evolution of physical variables *over time*, evolution in general relativity must be interpreted otherwise: not as the evolution of physical variables with respect to an independent variable 'time', but as *relative evolution of the physical variables with respect to one another*. Among the many physical variables, there are the proper times measured by clocks.

Example: Imagine you are holding two identical clocks. You keep one in your hand. You throw the other one upward: it goes up and then falls back down because of gravity and you catch it again (Figure 6.2). Since the two clocks have followed two different trajectories, they will measure different proper times. Let T_1 be the time measured by the hand-held clock and T_2 the time measured by the launched clock. Question: does the theory tell us how T_1 evolves as a function of real time T_2 or rather how T_2 evolves as a function of real time T_1? The question is meaningless. There is no 'real time'. The theory tells us the relationship between the two times.

Exercise: *Later we will calculate exactly the difference between T_1 and T_2 in the example. But the information given so far is sufficient to answer this question: which is greater between T_1 and T_2?*
Careful! *Special relativity and general relativity give opposite predictions!*
Hint: which clock follows a geodesic?

- ***Energy of the gravitational field***

Energy is the conserved quantity due to the invariance of a theory by time translations. General relativity is obviously invariant under translations in any time coordinate t. But due to the invariance for coordinate transformations, this translation is a local gauge symmetry. The generators of local gauge symmetries are always zero. It immediately follows that in general relativity the total energy, thus defined, is always zero.

This conclusion, surprising at first sight, also follows from a complete canonical analysis of the theory (which I do not present here): the Hamiltonian density of a general covariant theory vanishes identically on all solutions.

Another way of looking at the same result is to calculate the time-time component of Einstein's equations in the gauge $g_{00} = 1$, $g_{0i} = 0$. It is easy to see that the left-hand side of these equations has no second derivatives in time [show it!]. So this is not an evolution equation, but a constraint on the initial conditions, given by the fields and their first derivatives (like the corresponding first Maxwell equation). The right-hand side of the equation, however, is the energy density of matter. The left-hand term is a function of the metric and its first derivative which represents the energy density of the gravitational field (in this gauge) and is always exactly equal to and of opposite sign to the energy of matter, so that *the total energy density is zero at any point*.

Another way of looking at the same result is the following: we have seen that the energy-momentum tensor of matter, T_{ab}, is obtained as the variation of the action of matter with respect to g_{ab}. The *total* energy-momentum is the variation of the total action with respect to g_{ab}, and this is obviously zero: it is the equation of motion.

In particular conditions, it is possible to define specific notions of energy. For example, we can assign a *global* energy to an isolated system surrounded by a sufficiently flat space, taking as generators of time transformations the time translations in this flat space.

Similarly, weak gravitational waves can be seen as perturbations travelling over a Minkowski space, and we can assign an energy to them. But these and similar notions of energy work only in particular situations. In general, the notion of energy is simply not a meaningful notion for the gravitational field.

This is another radical consequence of the theory's background independence, that is, of the fact that in general relativity, physical events, described by fields and particles, are localised only with respect to each other.

The full consequences of this background independence will become clear in the following, using the theory explicitly. It's finally time to start doing it.

PART III **Applications**

7 Newtonian Limit

THE METRIC IN THE NEWTONIAN LIMIT

Let us consider a static physical situation in Newtonian physics where the Newtonian potential is weak. We want to see what this regime looks like in general relativity. Since everything is static, (we can find coordinates such that) the metric field is time independent. We can take inspiration from the Maxwell case, where the static field can be written (in a gauge) as containing only the time component of the Maxwell potential $A_a(\vec{x}) = (\Phi(\vec{x}), 0, 0, 0)$, and guess that this regime will be analogously given by the sole time-time component of $g_{ab}(x)$ differing from the Minkowski metric. This suggests considering the field

$$
g_{ab}(\vec{x}) = \begin{pmatrix} -(1 + 2\,\phi(\vec{x})) & 0 & 0 & 0 \\ 0 & 1 & 0 & 0 \\ 0 & 0 & 1 & 0 \\ 0 & 0 & 0 & 1 \end{pmatrix}.
\tag{7.1}
$$

(The reason for inserting the factor 2 will be clear soon.) As already mentioned, it is conventional to write a gravitational field using explicitly the form (3.17). Equation (7.1) can thus be written in the compact form

$$
ds^2 = -(1 + 2\phi(\vec{x}))\, dt^2 + dx^2 + dy^2 + dz^2.
\tag{7.2}
$$

Let's see how a particle moves in this metric if the particle is initially at rest in these coordinates. For a particle at rest, say at the origin, the world line is $x^a(\tau) = (\tau, 0, 0, 0)$, hence $\dot{x}^a = (1, 0, 0, 0)$. Therefore, the geodesic equation reduces to

$$
\ddot{x}^a + \Gamma^a_{oo} = 0.
\tag{7.3}
$$

Consider the space components of this equation, restricting the index a to be spacelike. Since the metric is time independent, the Levi-Civita connection reduces to

$$\Gamma^a_{oo} = \frac{1}{2}g^{ab}(-\partial_b g_{oo}) = -\eta^{ab}(\partial_b \phi), \qquad (7.4)$$

where in the last equality we have replaced the inverse metric with the Minkowski metric because we are working at first order around Minkowski in the weak field limit. This gives

$$\ddot{\vec{x}} = -\vec{\nabla}\phi. \qquad (7.5)$$

This is precisely the equation that gives the motion of a particle in a Newtonian potential $\phi(\vec{x})$! We conclude immediately that a particle in the metric (7.2) moves precisely like a particle in a Newtonian potential $\phi(\vec{x})$.

7.2   NEWTON'S FORCE

Let us now study conditions that the Einstein equations put on (7.2). If we assume a static energy momentum tensor formed by the sole component $T^{oo} = \rho$, where ρ is the matter density, and insert (7.2) into the Einstein equations, we obtain, with a bit of work, to linear order in the weak field ϕ, to the first relevant order in $1/c$, and disregarding the cosmological constant (which is small)

$$\Delta\phi = 4\pi G\rho, \qquad (7.6)$$

where $\Delta = \partial^2/\partial x^2 + \partial^2/\partial y^2 + \partial^2/\partial z^2$ is the Laplace operator. This is precisely the equation satisfied by the Newtonian potential! For a mass M concentrated at the origin, this gives the potential $\phi = -\frac{GM}{r}$ [show that this satisfies (7.6)] and hence the force (1.6) between two masses. Thus general relativity gives back the full universal gravitation, that is, Newtonian gravity, in the weak-field static limit.

Around a mass M, the Newtonian potential is $\phi = -\frac{GM}{r}$, where r is the distance from the mass. Therefore, in this approximation the spacetime geometry around a mass m is given by the metric

$$ds^2 = -\left(1 - \frac{2GM}{c^2 r}\right) c^2 dt^2 + dx^2 + dy^2 + dz^2, \qquad (7.7)$$

where I have reinstated $c \neq 1$ (fixed by dimensional analysis). On the surface of the Earth, $M = M_\oplus$ and $r = r_\oplus$, giving

$$\frac{2GM_\oplus}{c^2 r_\oplus} = \frac{2 \times 6.67 \times 10^{-8} \text{ cm}^3/\text{gs}^2 \times 5.972 \times 10^{24} \text{ kg}}{(3 \times 10^8 \text{ m/s})^2 \times 6{,}371 \text{ km}} \sim 1.3 \times 10^{-9}. \qquad (7.8)$$

Therefore, the correction to the Minkowski metric around us is of the order of one part in a billion. As the equations in this paragraph show, this is sufficient for causing things to fall down: the geodesics of such a slightly altered metric are the motions of the falling bodies.

7.3   GENERAL RELATIVISTIC TIME DILATION

So far, we have derived known physics from a limit of general relativity. Let's move to the first genuinely new prediction of the theory.

On the surface of the Earth, the gravitational potential is $\phi = gh$, where $g \sim 9.8$ m/s^2 and h is the vertical elevation. Hence the metric is, approximately,

$$ds^2 = -(1 + 2gh)dt^2 + dx^2 + dy^2 + dz^2. \qquad (7.9)$$

Take two equal clocks. Keep one at ground level and raise the second to the altitude h for a coordinate time t. Then bring it back to ground level and compare it with the other clock. See Figure 7.1. During the interval t, the proper time T_{down} measured by the ground clock is t, because at $h = 0$ the metric is Minkowski. But not so for the upper clock. Here, the proper time is

$$T_{up} = \int_0^t \sqrt{(1 + 2gh)dt^2} \sim (1 + gh)\, t > T_{down}. \qquad (7.10)$$

This is a rather spectacular result: a clock runs faster if it is at higher altitude. The fractional time difference, restoring $c \neq 1$ using

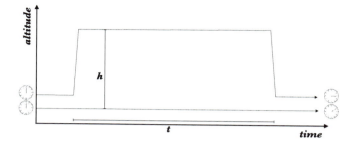

FIGURE 7.1 Time passes faster in altitude: here, two equal clocks are separated and kept at different altitudes. When back together, the lower one is late.

dimensional analysis, is

$$\frac{\Delta T}{T} = \frac{T_{up} - T_{down}}{T_{down}} = \frac{gh}{c^2}. \tag{7.11}$$

At $h = 1$ m, this gives

$$\frac{\Delta T}{T} = \frac{9.8 \text{ m/s}^2 \times 1 \text{ m}}{(3 \times 10^8 \text{ m/s})^2} \sim 10^{-16}. \tag{7.12}$$

This means that if a clock is kept 1 metre higher than another for 100 days ($\sim 10^7$ s), the lower clock remains late by ~ 1 ns.

Today's best clocks have a precision superior to 10^{-16} and therefore this effect can be detected in the laboratory. It has indeed been verified: time runs faster for your head than for your feet.

Einstein's intuition that gravity can be understood as a modification of the spacetime geometry leads to an immediate physical prediction, which today we can verify in the lab: clocks run slower if they are lower in the gravitational potential.

- **_Newtonian gravity as an effect of time dilatation_**

A moment of reflection on the results of Sections 7.1 and 7.3 leads to a rather spectacular realisation. Masses subjected to the gravitational force follow geodesics in a spacetime whose only difference from Minkowski space is a local modification of g_{00}, that is, a modification of the local speed at which time passes (compared to the speed at which it passes elsewhere). Therefore, in general relativity we can say that things fall towards a mass *because* a mass slows down time in its vicinity.

- **Size of the gravitational effects**
If the correction to the Minkowski metric is of order 10^{-9} (equation (7.8)), why is the effect on the clocks only of order 10^{-16} (equation (7.12))? The reason has to do with the freedom of choosing coordinates. If the correction $\phi(\bar{x})$ to the Minkowski metric was constant in space, it would be a trivial change of coordinates: no physical effect. What matters here is the gradient of $\phi(\bar{x})$, which multiplied by the displacement h considered gives the smaller number.

- **Exercise: *Throwing a clock upward***
Consider Figure 6.2. A clock is thrown upward with an initial velocity v_0, and its reading is compared with the reading of a clock that has remained in your hand. Compute the motion of the clock using elementary Newtonian mechanics and the difference between the times measured by the two clocks (i) according to special relativity and (ii) according to general relativity. Why do you get opposite signs? Which is right?

- **Exercise: *The GPS***
The Global Positioning System works with precision clocks mounted on satellites in orbit around the Earth. A clock on a satellite runs faster than clocks on Earth because of the general relativistic time dilatation studied in this chapter. But since a clock in orbit is moving with respect to Earth, it runs slower because of Lorentz time dilatation. The two effects have opposite sign and have a different dependence on the altitude. There could therefore be an altitude where the two effects cancel. Find this altitude.
    Solution. *A satellite at a distance r from the centre of the Earth has a gravitational acceleration $a = GM/r^2$. Assuming it is in a circular orbit, this is balanced by the centrifugal acceleration $a = v^2/r$. Hence $GM/r^2 = v^2/r$ and the speed at altitude r is $v^2 = GM/r$. The special relativistic time dilatation is given by (1.4), which gives*

$$\Delta T_{SR}/T = 1 - \gamma = 1 - \sqrt{1 - v^2/c^2} \sim \frac{1}{2}\frac{v^2}{c^2} = \frac{1}{2}\frac{GM}{rc^2}. \qquad (7.13)$$

The general relativistic time dilatation is not given by (7.10) because a satellite is outside the approximation where the Newtonian potential is gh. The Newtonian potential is rather $\phi = GM/r$. Therefore, (7.10) becomes

$$T_{up} = \int_0^t \sqrt{(1 + 2\phi)dt^2} \sim (1 + \phi)t = (1 - \frac{GM}{r})t > T_{down} = (1 - \frac{GM}{R})t$$

$$(7.14)$$

where R is the altitude of the lower clock, which is the radius of the Earth. It follows that

$$\Delta T_{GR}/T = \frac{T_{up} - T_{down}}{T_{down}} = \frac{GM}{rc^2} - \frac{GM}{Rc^2}. \tag{7.15}$$

The two time dilatations balance when

$$-\frac{1}{2}\frac{GM}{rc^2} = \frac{GM}{rc^2} - \frac{GM}{Rc^2}, \tag{7.16}$$

which gives

$$r = \frac{3}{2}R. \tag{7.17}$$

Since the radius of the Earth is $\sim 6,000$ km, the altitude of the orbit where the time dilatation vanishes is $h = r - R \sim 3,000$ km. At lower orbits the orbiting speed is higher, hence the special relativistic effect is stronger, while the general relativistic effect is weaker because the potential difference is less. Hence clocks run slower. At higher orbits, they run faster.

- **Exercise: GPS without GR**
The satellites of the American GPS system orbit at radius $R \sim 20,000$ km. Compute how long it would have taken the system to accumulate a localisation error of 3 km on Earth if the system had been put in place before the discovery of general relativity, hence ignoring the general relativistic effect.
    Solution. GPS signals travel at the speed of light, hence they cover the length $l = 3$ km in a time $\Delta T = l/c \sim 10^{-5}$ s. The relativistic correction is

$$\Delta T/T = \frac{GM}{R_{\oplus}c^2} - \frac{GM}{Rc^2} \sim 10^{-9}. \tag{7.18}$$

Hence $T = \Delta T/10^{-9} \sim 10^4$ s: less than three hours. The understanding of general relativity has played a key role in the construction of navigation systems such as GPS.

- **Exercise: The clock at the centre of the Earth**
A clock is placed at the centre of the Earth. Compute how much it will be late with respect to a clock on the surface of the Earth after one year.

8   Gravitational Waves

The previous chapter studied the Newtonian limit of GR, which corresponds to the Coulomb solution of the Maxwell equations. Here we study gravitational waves, which correspond to the electromagnetic wave solution of the Maxwell equations.

Einstein got confused about the reality of gravitational waves. He first thought that they are real, by analogy with electromagnetism. Later he changed his mind and convinced himself that there are no gravitational waves (we shall see why). Then he changed his mind again and decided that they are real after all. The scientific community as well was long confused in the following decades. The discussion was settled in the 1960s, thanks mostly to the works of Felix Pirani and Hermann Bondi.[1] Indirect evidence for the reality of these waves grew steadily in the 1970s. The search to detect them directly started shortly after and took more than thirty years.

The direct detection of gravitational waves was finally accomplished in 2015,[2] and led to the Nobel Prize in 2017, crowning a century-long saga.

- **Electromagnetic waves**

Let us briefly review the theory of electromagnetic waves, which is a good model for the gravitational waves. The electromagnetic field is described by the potential A_a, and its field equations in the absence of charges are

$$\partial_a F^{ab} = 0, \tag{8.1}$$

---

[1] H. Bondi, F. A. E. Pirani, I. Robinson (1959). 'Gravitational waves in General Relativity III: Exact plane waves'. *Proc. Roy. Soc. A.* 251 (1267): 519–533.
[2] B. P. Abbott, *et al.* (LIGO Scientific Collaboration and Virgo Collaboration) (2016). 'Observation of gravitational waves from a binary black hole merger'. *Phys. Rev. Lett.* 116 (6): 061102.

where $F_{ab} = \partial_a A_b - \partial_b A_a$. Inserting this into the equation gives

$$\partial^a \partial_a A_b - \partial_b \partial_a A^a = 0. \tag{8.2}$$

This equation can be simplified using the gauge invariance of the theory. The fields A_a and

$$\tilde{A}_a = A_a + \partial_a \lambda \tag{8.3}$$

are gauge equivalent. By selecting λ appropriately we can therefore choose a gauge where $\partial_a A^a = 0$ and the equation reduces to

$$\partial^a \partial_a A_b = 0 \tag{8.4}$$

which is the wave equation for each component of $A_b = (A_o, \vec{A})$. We have not exploited the gauge invariance entirely, because we can still fix $A_o = 0$, which together with the gauge condition gives $div\vec{A} = 0$. Therefore, the relevant equations are

$$\partial^a \partial_a \vec{A} = 0, \qquad div\vec{A} = 0. \tag{8.5}$$

The first is solved by any linear combination of plane waves of the form (real part of)

$$\vec{A}(\vec{x}, t) = \vec{\epsilon} \, e^{i(\vec{k}\cdot\vec{x} - \omega t)}, \tag{8.6}$$

where $\vec{\epsilon}$ is a constant polarisation vector and provided that $\omega^2 = |k|^2$. The second implies that

$$\vec{\epsilon} \cdot \vec{k} = 0, \tag{8.7}$$

namely the polarisation of the wave is transverse with respect to the direction of propagation. For instance, a wave travelling in the z direction has two components:

$$\vec{A}(\vec{x}, t) = (\epsilon_x, \epsilon_y, 0) \sin(k(z - t)). \tag{8.8}$$

Notice that the wave comes back to itself if we rotate it by an angle π along the axis of propagation. This is a way of saying that it has spin one.

We can now return to gravity, using electromagnetic (EM) waves as a model.

- **Linear gravitational waves**

Plane waves of small amplitude are predicted by the Einstein equations. They are small ripples of the geometry perturbing flat spacetime. Let us assume that the metric is nearly flat. Then we can write it as

$$g_{ab}(x) = \eta_{ab} + h_{ab}(x), \tag{8.9}$$

where η is the Minkowski metric and the entries of the matrix h_{ab} are small compared to unit. From now on, we disregard terms of quadratic or higher order in h_{ab}.

The gauge transformations (6.1) maintain the form (8.9) if $\tilde{x}^a(x) = x^a + \lambda^a(x)$, where λ is small; then to first order in λ the metric transforms under this gauge transformation as

$$\tilde{h}_{ab} = h_{ab} + \partial_a \lambda_b + \partial_b \lambda_a. \tag{8.10}$$

(Compare with (8.3).) It is not hard to show that, using this gauge transformation, it is possible to bring h_{ab} to satisfy the three gauge conditions

$$\partial^a h_{ab} = 0, \qquad \eta^{ab} h_{ab} = 0, \qquad h_{oa} = 0. \tag{8.11}$$

If we insert (8.9) into the vacuum Einstein equations and we discard all quadratic terms in h_{ab}, we obtain a linear second order partial differential equation. To linear order in h, we have

$$\Gamma^a_{bc} = \frac{1}{2}\eta^{ad}(\partial_b h_{dc} + \partial_c h_{db} - \partial_d h_{bc}). \tag{8.12}$$

The Riemann tensor reduces to

$$R^a{}_{bec} = \partial_e \Gamma^a_{bc} - (e \leftrightarrow c) = \frac{1}{2}\eta^{ad}(\partial_e \partial_b h_{dc} + \partial_e \partial_c h_{db} - \partial_e \partial_d h_{bc}) - (e \leftrightarrow c) \tag{8.13}$$

so that the vacuum equations without cosmological constant $R_{ab} = 0$ are

$$R^a{}_{bac} = \frac{1}{2}(\partial^d \partial_b h_{dc} - \partial^d \partial_d h_{bc} - \eta^{da} \partial_c \partial_b h_{da} + \partial_c \partial^d h_{bd}) \quad (8.14)$$

which, using the gauge condition, reduces to the wave equation

$$\partial^d \partial_d h_{ab} = 0. \quad (8.15)$$

This is solved by linear combinations of plane waves

$$h_{ab}(\vec{x}, t) = \epsilon_{ab} \, e^{i(\vec{k} \cdot \vec{x} - \omega t)}, \quad (8.16)$$

where, as above, $\omega^2 = |k|^2$, while ϵ_{ab} is symmetric, has only space components, is traceless $(\eta^{ab}\epsilon_{ab} = 0)$, and is transverse $(k^a \epsilon_{ab} = 0)$. For instance, a wave travelling in the z direction has the form

$$h_{ab}(\vec{x}, t) = \begin{pmatrix} 0 & 0 & 0 & 0 \\ 0 & \epsilon_+ & \epsilon_\times & 0 \\ 0 & \epsilon_\times & -\epsilon_+ & 0 \\ 0 & 0 & 0 & 0 \end{pmatrix} \sin(k(z - t)). \quad (8.17)$$

The line element of the two polarisations is therefore

$$ds^2 = -dt^2 + \left(1 + \epsilon_+ \sin(k(z-t))\right) dx^2 + \left(1 - \epsilon_+ \sin(k(z-t))\right) dy^2 + dz^2 \quad (8.18)$$

and

$$ds^2 = -dt^2 + dx^2 + dy^2 + dz^2 + 2\epsilon_\times \sin(k(z - t))dxdy. \quad (8.19)$$

These are two plane gravitational waves. A moment of reflection shows that the second is just the first rotated by 45 degrees around the z axis, and each of them transforms back into itself for a rotation of $\pi/2$. This is a way of saying that these waves are spin two.

Notice that while (8.8) are *exact* solutions of the Maxwell equations, these waves are only approximate solutions of the Einstein equations. This implies that (8.18) and (8.19) describe real waves only if the (dimensionless) amplitudes $\epsilon_\times$ and ϵ_+ are small. If the amplitudes are large, non-negligible non-linear effects set them.

8.1   EFFECT ON MATTER

If a plane electromagnetic wave arrives where there is a charge, the charge oscillates, moved up and down by the oscillating electric field. What happens if a gravitational wave arrives where there is a mass? Let us compute this, and be ready for a surprise.

Take a mass that is initially not moving, at the origin. Hence $\dot{x}^a = (1, 0, 0, 0)$. Its geodesic equation is, therefore,

$$\ddot{x}^a + \Gamma^a_{oo} = 0. \tag{8.20}$$

It is immediately clear that Γ^a_{oo} vanishes for a gravitational wave, because all h_{oa} components of h vanish. Therefore $\ddot{x}^a = 0$, the particle does not move! This is surprising and, at first, confusing: what is a gravitational wave if it does not move a mass?

It is not strange that everybody was confused by gravitational waves for so long!

● *The solution of the puzzle*
The solution of the puzzle is to recall that coordinates do not mean anything in general relativity. The fact that the coordinates do not change does not mean anything. Any moving object can be described by coordinates that do not change: it suffices to use the object itself to define the coordinates.

To understand what happens, we must consider *two* masses. Suppose we have one mass at the origin $(x = y = z = 0)$ and one mass at the coordinates $y = z = 0$ but $x = L \neq 0$. Then both masses do not move in these coordinates, but their distance changes because the distance is given by

$$D = \int_o^L \sqrt{g_{xx}}\, dx = \sqrt{1 + \epsilon_+ \sin(k(z - t))}L \sim L + \frac{L}{2}\epsilon_+ \, \sin(\omega t). \tag{8.21}$$

The physical and geometrical distance between the two particles oscillates with time!

Consider, for instance, a rod and two masses free to move near the rod. The rod is (approximately) rigid, therefore (tidal) gravitational stresses do not deform it much. The distance between its two extremes remains the same. But the masses are moved apart by gravity and they move with respect to the rod.

Is it the rod that stretches and compresses, while the masses do not move, or does the rod stay put and the masses move?

The question is meaningless!

There is no notion of 'moving' in general relativity, unless we refer to something else! The two images in Figure 8.1 are two ways to describe the same physical reality: the rod and the masses move *with respect to one another*. The difference is only the way in which the motion is represented with respect to the background of the page of the book: in nature, there is no similar background.

• *The logic of the current gravitational antennas*

Equivalently, imagine that we send light pulses from one mass to the second, where there is a mirror so that the pulses bounce back. Imagine we time the back-and-forth travel time of the pulses (by measuring the proper time on one mass, from the departure to the

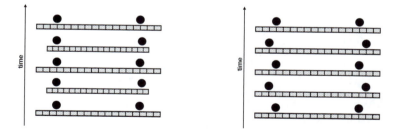

FIGURE 8.1 The effect of a gravitational wave on two masses near a rod. Left: the masses do not move, and the rod, which measures the distance, oscillates. Right: the rod keeps its shape, and the position of the masses oscillates. These are two pictures representing *the same* physical reality: the motion of the masses relative to the rod. Study the picture with care: if you understand it clearly, then you understand background independence clearly.

return of the light pulse). From the light ray equation $ds = 0$ we see that, in the coordinates we are using, the speed of light varies with time:

$$dx/dt = 1 - \frac{1}{2}\epsilon_+ \sin(k(z - t)). \qquad (8.22)$$

Hence we will see that the travel time oscillates. This does not depend on the coordinates used. We shall see soon that this is the way gravitational waves are measured.

- **Action of a wave on a circle of masses**

Finally, notice that in (8.18) the change of g_{xx} and the change of g_{yy} are in counter-phase (because of the minus sign in the second). This means that when the x direction shrinks, the y direction expands, and vice versa. A ring of particles in the $x-y$ plane would move (with respect to a rigid desk) as in Figure 8.2 when a gravitational wave falling from the z direction hits them: it will produce an oval that oscillates in counter-phase in the perpendicular directions. The effect of the wave (8.19) is the same rotated by 45 degrees [*show it!*].

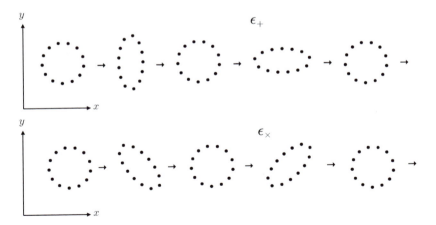

FIGURE 8.2 The motion of a ring of particles under the effect of a gravitational wave travelling perpendicularly to the plane of the paper. Above: the + polarisation. Below: the × polarisation. Notice the two are rotated 45 degrees with respect to one another.

8.2   PRODUCTION AND DETECTION

• *Electromagnetic dipole*

Recall that to produce an *electromagnetic* wave we need a distribution of charge varying in time. An isolated distribution of charges $\rho(x, t)$ can be expanded in a multipole expansion. The lowest terms are the total charge

$$q(t) = \int \rho(x, t)\, d^3x \qquad (8.23)$$

and the dipole

$$d^i(t) = \int \rho(x, t) x^i\, d^3x. \qquad (8.24)$$

A time variation of $q(t)$ would produce a spherically symmetric wave, but charge conservation forbids $q(t)$ to vary. Therefore, there are no spherically symmetric EM waves, and the lowest-order electromagnetic waves are dipolar. They are generated by a dipole $d^i(t)$ varying in time. An oscillating dipole is, for instance, produced by charges moving up and down in an antenna.

• *Gravitational quadrupole*

A *gravitational* wave is produced by a varying distribution of mass-energy. A distribution of energy $\rho(x, t)$ can be expanded in a multipole expansion. The lowest terms are the total energy and the dipole. Energy conservation forbids total energy to vary in time. But momentum conservation forbids the dipole to vary as well, because the time derivative of the dipole is just the total momentum. Hence gravitational waves are only generated by the next order of the expansion: the quadrupole

$$q^{ij}(t) = \int \rho(x, t) x^i x^j\, d^3x. \qquad (8.25)$$

An oscillating quadrupole is produced, for instance, by two masses connected by an oscillating spring, or by two masses orbiting around each other. The emitted wave has a quadrupolar structure, which is reflected in the way the wave affects masses, as displayed in Figure 8.2: notice in fact that the dipole of a group of masses moving as shown in the figure is constant.

Question (difficult): Two small masses orbiting each other follow geodesics. In its local frame of reference, each is in free fall, has zero acceleration, and so emits no radiation. Why do the two masses emit radiation? This is one of the initial sources of confusion about gravitational emission. Hint: draw the Faraday lines of an electric charge falling on the surface of the Earth, taking into account the finite speed of information propagation.

From the linearised version of his equations, Einstein derived a formula, analogous to a similar electromagnetic one, for the value of the perturbation h_{ij} of the Minkowski metric at a distance r from a varying quadrupole. This is called Einstein's quadrupole formula:

$$h_{ij}(r, t) = \frac{2G}{rc^4} \ddot{I}_{ij}(t - r/c),$$ (8.26)

where I have restored physical units, the two dots indicate the second time derivative, and $I_{ij} = q_{ij} - \frac{1}{3}\delta_{ij}\delta_{lm}q^{lm}$. I do not give the derivation of this formula. From this it is not difficult to show that the power radiated by a time-varying energy distribution, namely its gravitational luminosity, is

$$P = \frac{G}{5c^5}\langle \dddot{I}_{ij} \dddot{I}^{ij}\rangle,$$ (8.27)

where the angle brackets indicate time averaging.

The factor $\frac{G}{5c^5}$ is extremely small. This is why, to produce significant gravitational radiation, extremely highly relativistic systems are needed.

The only gravitational waves detected so far have been produced by the final moments of the merging of black holes and neutron stars.

• *Black holes merger*

Two objects orbiting around each other form a variable quadrupole and thus emit gravitational waves. If the two objects are very massive, close, and fast, the emission becomes considerable. Due to the loss of energy into the emission of gravitational waves, the two objects lose energy and therefore get closer. Their orbit becomes a spiral: they fall towards each other.

The final stages of this process are catastrophic, because as the energy decreases, the radius of the Keplerian orbits decreases, and their frequency and velocity increase. The result is that the emission of gravitational waves increases. There is a final, violent explosion after which the two objects collide.

The shape of the waves emitted by this process is characteristic. A wave progressively increases in frequency and intensity, giving a characteristic 'chirp': a rapid final increase in frequency and amplitude.

- ### Detection

Current gravitational wave detectors compare the distances between a central point and two masses, placed at the end of the two arms at 90 degrees from each other. The masses are hung and therefore free to move in the horizontal plane. When a wave arrives from the vertical, the length of the two arms oscillates in counter-phase, like four masses perpendicular to each other in Figure 8.2.

The difficulty of the measurement is due to the smallness of the amplitude of the waves arriving on the Earth: $h \sim 10^{-21}$. That is, we need to measure the change of a length in one part in 10^{21}. This is done with an interferometer, which compares the phases of two laser beams travelling along the two arms, each between the centre point and a mass. When a gravitational wave passes, the two axes of the interferometer change length in counter-phase. The interferometer measures the difference in the relative phase, which oscillates with the frequency of the wave. The LIGO detector is formed by two such interferometers (restricting to coincident detections allows the filtering of noise).

On 14 September 2015, both LIGO interferometers picked up a signal lasting 0.2 second, produced by two black holes, respectively of about 30 and 35 solar masses, which merged into a single

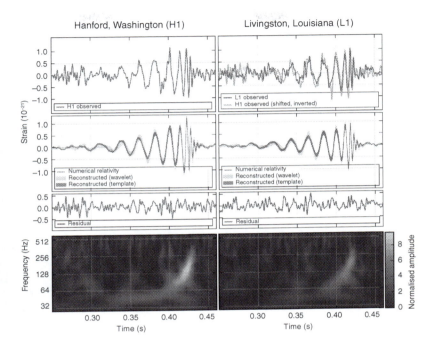

FIGURE 8.3 The signal of the first gravitational wave measured by the two LIGO detectors. The image shows the detected waveform ('Strain' is the relevant component of h_{ij}); the expected waveform, obtained with numerical calculations; and a representation of the change in frequency and pitch. Notice the characteristic 'chirp'. (B. P. Abbott *et al.* – LIGO Scientific Collaboration and Virgo Collaboration; CC BY 3.0)

black hole of 62 solar masses. *[Where did the missing mass go?]* The frequency of the observed signal, which corresponds to twice the orbital frequency *[why the factor 2?]*, increased from 35 Hz to 250 Hz during the 0.2 second. The two black holes orbited at $\sim$ 350 km of each other, and their orbital speed increased from 30 to 60% of the speed of light.

The merger took place about a billion years ago (in cosmological time; see Chapter 9 for the definition of cosmological time): the gravitational wave has been travelling ever since. The power emitted by the gravitational waves during the last milliseconds of the merger was 50 times greater than the combined power of all the light radiated by all the stars in the observable universe.

Exercise: *From the data given, estimate the order of magnitude of the quadrupole of the two black holes and its second derivative (from the frequency) and use equation (8.26) to estimate the amplitude of the wave that reaches the Earth. Compare with the data measured by LIGO, shown in Figure 8.3. Do the numbers match?*

Question: *Where was the enormous energy radiated in gravitational waves taken from?*

9   Cosmology

Modern cosmology is the study of the history of the very large structure of the universe. It is a successful and booming field: in the last few decades we have been able to credibly reconstruct the evolution of the large-scale degrees of freedom of the universe we observe, as far back as approximately 13 billion years into the past.

The field was started by a remarkable paper that Albert Einstein published in 1917. The paper single-handedly jump-started modern cosmology.[1] The paper is amazing for its incredible mixture of astonishing intuitions and ideas and glaring mistakes.

- *Astonishing intuitions and ideas...*

The first idea is to study the large-scale structure of the universe. The correct intuition here is that this structure is largely dominated by gravity.

The second idea is that the universe could be well approximated by considering a *uniform* distribution of matter. Matter is obviously *not* uniformly distributed in the universe (there are stars and empty interstellar spaces, galaxies and empty intergalactic spaces, and galaxy clusters and empty spaces in between). At the time there was no indication of uniformity at large scales. But Einstein's intuition turned out to be correct: there is uniformity at larger scales, and the homogeneous model provides a decent first-order approximation to the dynamics of the universe.

The third great idea by Einstein is that since the geometry of space is curved, the universe can have finite volume and no boundaries. There are, in fact, Riemannian geometries that have finite

---

[1] A. Einstein (1917). 'Kosmologische Betrachtungen zur allgemeinen Relativitätstheorie', *Sitzungsberichte der Preußischen Akademie der Wissenschaften*: 142.

volume and no boundaries. The obvious example in two dimensions is the sphere. The analogous geometry in three dimensions is the 3-sphere, which we have already studied in Section 3.2.

On the basis of these notions, Einstein's idea was to treat the scale of the homogeneous geometry of space (the radius $a(t)$ of the sphere) as a dynamical quantity and use the equations of general relativity to study its dynamics: the dynamics of the largest degree of freedom of the universe.

All this has turned out to be extremely farsighted. But there is much more in that paper.

- *... glaring mistakes ...*

There are two mistakes in the paper, of different natures.

An immediate application of his equations to a uniform universe as a whole indicates that if the cosmological constant is zero, the universe cannot remain stationary: it either contracts or expands. The reason is not difficult to see. It is the same reason that a stone cannot stand still in midair: it either falls or (if it has been thrown) moves upwards.

In 1917, the expansion of the universe had not yet been observed, nor was it suspected. Einstein's theory predicted it. Einstein should have stopped there and published a spectacular prediction, to be confirmed about a decade later: the universe is not static on a large scale.

Instead, he makes his first mistake: he does not believe in his own theory and therefore loses the opportunity for a great correct prediction. It is said that he later regretted this mistake as the biggest one of his life.

Instead of believing his theory, he trys to patch it up. And here, he makes his second mistake, even more incredible. To tweak his theory and force it to agree with the stationarity of the universe, which he (erroneously) assumed, Einstein introduces the cosmological term $+\lambda g_{ab}$ into equation (4.12). The rationale is that if matter density and λ are properly calibrated, the equations admit a stationary solution.

The cosmological term (as we will see in detail in Section 10.5), indeed, gives rise to a repulsive force that can balance the Newtonian attraction. But there is a problem: Einstein fails to notice that this equilibrium is unstable. The universe cannot be described by this solution because fluctuations and non-homogeneity would have already taken it away from the stable solution.

The reason for the instability is so simple that it is surprising that Einstein did not notice it: the Newtonian force becomes stronger as the masses approach each other, while the cosmological repulsion becomes stronger as they move away. At equilibrium they can balance, but a minimum approach brings them closer, while a minimum distancing makes them move further away: the equilibrium is unstable. With or without the cosmological constant, general relativity predicts that the universe is expanding or contracting, and Einstein didn't notice. How could the great Einstein not notice this obvious instability? Yet it happened. Maybe he was in love or distracted.

- ## ...and more spectacular insights

And yet – despite the fact that it doesn't make the universe static as he would have liked – the cosmological term that Einstein introduces in the equations ends up being another stroke of genius. The term is right! The fact that the cosmological constant is not zero has now been measured, more than half a century later!

So Einstein correctly guessed the existence of the cosmological force, which we now know really exists. But he did it for the wrong reason: he thought this would result in a static universe, and that's not true. So he missed a most spectacular prediction of his own theory: the universe cannot be static at large scale. But he ended up introducing a correct improvement of his equations.

Formidable.

- ## Lemaître

The first to understand the full cosmological implications of general relativity and to glimpse evidence of the expansion in the astronomical data on the red shift of the galaxies (called 'nebulae' at the time) was Georges Lemaître.

Today the evidence for the expansion is overwhelming.

From the observed expansion rate, the equations of general relativity allow us to compute the age of the universe and its full past evolution since an initial very small and compressed state that Lemaître called the 'primordial atom' and we call today the 'Big Bang'.

Lemaître realised also that the physics of this primordial atom was likely to involve quantum effects. Today it is a common opinion that studying what happened around the Big Bang requires quantum gravity. I will touch upon this issue in the last chapter. For now, let us describe the large-scale geometry of the universe and the equations governing it.

9.1   THE LARGE-SCALE GEOMETRY OF THE UNIVERSE

In Section 3.2 we derived the metric (3.50) of a 3-sphere of size a. If a depends on t and we add the (proper) time coordinate t, we obtain the Lorentzian 4d metric

$$ds^2 = -dt^2 + a^2(t) \left(\frac{dr^2}{1 - r^2} + r^2 d\Omega^2 \right). \tag{9.1}$$

This is the metric studied by Einstein in his 1917 paper. It is determined by a single function $a(t)$. The spacetime it describes is a homogeneous and isotropic 3d space of radius a which varies in time t. For a mass that is stationary in these coordinates the time t is equal to the proper time. So clocks staying put in these coordinates (called 'co-moving') remain synchronised and measure a common proper time, which is called the 'cosmological time'. The possibility of defining such a common time is lost if we leave the homogeneity approximation. Every galaxy is stationary in these coordinates. (Obviously this is an approximation: Andromeda will collide with the

Milky Way soon.) What changes during the expansion is the distance between galaxies, given by a change in the metric.

More generally, we can consider the time-dependent metric of a generic homogeneous 3d space:

$$ds^2 = -dt^2 + a^2(t)\left(\frac{dr^2}{1-kr^2} + r^2 d\Omega^2\right),\qquad(9.2)$$

where $k = 0, \pm 1$.

The actual size of the universe is unknown: there is observational evidence that it is quite larger than the part we can observe. It turns out that the approximation of the 3-sphere with a flat 3-geometry is good enough in current cosmology and therefore we often use $k = 0$. This means that space (but not spacetime) is flat.

• **Hubble's law**

Assuming $k = 0$, the distance between two galaxies at coordinate distance Δr is

$$D = a(t)\Delta r.\qquad(9.3)$$

If the universe is expanding, the relative velocity between two galaxies at distance D is

$$V = \frac{dD}{dt} = \dot{a}(t)\Delta r = \frac{\dot{a}(t)}{a(t)}D.\qquad(9.4)$$

Therefore, the ratio V/D between the velocity and the distance between any two galaxies is independent from the distance between them. The proportionality constant

$$H = \frac{\dot{a}}{a}\qquad(9.5)$$

is called the Hubble constant, after the astronomer Edwin Hubble. It is a quantity directly measurable because the relative velocity of galaxies with respect to us can be evaluated from the Doppler shift of their spectral lines, while the distance can be determined by a number of indirect methods. (These represent a remarkable success of observational astronomy.) The current measured value of the Hubble constant is

$$H \sim 72 \,(\text{km/s})/\text{Mpc}, \tag{9.6}$$

given in the peculiar units astronomers like: kilometres per second, per mega-parsec. (A parsec is $\sim 3 \times 10^{10}$ cm, or ~ 3.2 light years. It is the distance of a star whose parallax, namely its apparent movement due to the orbit of the Earth, is one second.) This is the currently measured speed of expansion of the universe. The relation

$$V = HD \tag{9.7}$$

is called Hubble's law. The point of the 'law' is that H is independent from the distance of the galaxy observed.

• **The Friedmann equation**

Assuming that matter is uniformly distributed with density $\rho(t)$ and stationary in the co-moving coordinates, and inserting the metric (9.2) into the Einstein equations, we obtain the following differential equation for $a(t)$:

$$\frac{\dot{a}^2}{a^2} + \frac{k}{a^2} - \frac{\lambda}{3} = \frac{8}{3}\pi G\rho. \tag{9.8}$$

This equation was derived by Alexander Friedmann and is called the Friedmann equation.[2] It governs the large-scale dynamics of the geometry of the universe.

• **The energy density**

To solve the Friedmann equation we must know how the energy density ρ changes with the scale factor, namely with the expansion of the universe. This can be determined from the pressure terms in T_{ab} and other components of the Einstein equations, or more directly as follows.

---

[2] A. Friedmann (1922). 'Über die Krümmung des Raumes.' *Zeitschrift für Physik* 10 (1): 377–386. English translation in: Friedmann, A. (1999). 'On the curvature of space'. *General Relativity and Gravitation* 31 (12): 1991–2000.

For incoherent matter, the total energy ρV in a coordinate region of volume $V \sim a^3$ remains constant when V varies; therefore, $\rho = \rho_m / a^3$ where ρ_m is constant in time.

For electromagnetic radiation, there is an additional effect. The expansion of spacetime stretches the EM waves. A simple shortcut to see how the electromagnetic energy evolves is to consider that in the expansion the number of photons remains constant in a coordinate region when the volume increases, but each photon has energy $E = h\nu$ where ν is the frequency, which scales like $1/a$ in the expansion. Therefore, for radiation there is an additional $1/a$ effect; that is, $\rho = \rho_g / a^4$, where ρ_g is constant in time. Bringing it all together, the Friedmann equation reads

$$\frac{\dot{a}^2}{a^2} + \frac{kc^2}{a^2} - \frac{\lambda}{3} = \frac{8}{3}\pi G \left(\frac{\rho_m}{a^3} + \frac{\rho_g}{a^4} \right). \tag{9.9}$$

Taking the time derivative of this equation gives

$$\ddot{a} = -\frac{4}{3}\pi G \left(\frac{2\rho_g}{a^3} + \frac{\rho_m}{a^2} \right) + \frac{\lambda}{3}a. \tag{9.10}$$

The first term on the right-hand side is negative: it makes the expansion decelerate. The second is positive; it makes the universe accelerate. As long as a is small enough, the first term dominates and the universe decelerates, pulled by gravity. If a becomes sufficiently large, the second term wins and the expansion of the universe accelerates.

- **The age of the universe**

Astrophysical data indicate that for most of the past life of the universe, a was sufficiently small for the deceleration to dominate. This implies that the expansion rate has been lower than today in the past, and therefore the universe cannot have had a life longer than $H^{-1} = a/\dot{a}$, giving the value of H in years:

$$T_H < \frac{1}{H} \sim 14 \text{ billion years.} \tag{9.11}$$

Because of the different powers of the scale factor in the Friedmann equation, the expansion of the universe was dominated by the radiation in an early phase, then by matter, and later on by the cosmological constant. We are now in a matter-dominated expansion phase, but the effect of the cosmological constant is already detectable and has been detected.

- **Flatness**

The data indicate that the k term is small. This means that the universe is much larger than the part that we directly observe. This does not necessarily mean that $k = 0$ and that the universe is spatially flat: deducing that the universe is flat by the current smallness of this term is to make the same mistake as deducing that the Earth is flat just because we fail to detect its curvature at our usual scale.

9.2 BASIC COSMOLOGICAL MODELS

- **Matter- and radiation-dominated expansion**

In the regime where the cosmological, radiation, and k terms are small, and the only relevant source is matter, the Friedmann equation reduces to

$$\frac{\dot{a}^2}{a^2} = \frac{8}{3}\pi G \, \frac{\rho_o^m}{a^3}, \tag{9.12}$$

which is solved by

$$a(t) = a_o \, t^{\frac{2}{3}}. \tag{9.13}$$

This is a decelerating expansion. The constant a_0 can be chosen arbitrarily by rescaling the spatial coordinates. It is conventional to use co-moving coordinates that give physical distance now. So that $a(now) = 1$.

Exercise: Show that for a radiation-dominated universe $a(t) = a_o \, t^{\frac{1}{2}}$.

- **De Sitter universe**

When instead, we can disregard all the terms except for the cosmological one, the equation becomes

$$\dot{a} = \sqrt{\frac{\lambda}{3}} \, a \tag{9.14}$$

which is easily solved by

$$a(t) = a_0 \, e^{\sqrt{\frac{\lambda}{3}}\, t}. \tag{9.15}$$

This cosmological solution was found by Willem de Sitter and is called the de Sitter spacetime.[3] The line element of the de Sitter spacetime is

$$ds^2 = -dt^2 + e^{2\sqrt{\frac{\lambda}{3}}\, t}(dr^2 + r^2 d\Omega^2) \tag{9.16}$$

in co-moving coordinates. Light rays are governed by $ds = 0$; therefore, the speed of light in these coordinates is

$$dr/dt = e^{-\sqrt{\frac{\lambda}{3}}\, t}. \tag{9.17}$$

A light ray emitted at the origin at time t arrives at time $+\infty$ at the radius

$$r = \int_t^\infty e^{-\sqrt{\frac{\lambda}{3}}\, t} dt = \sqrt{\frac{3}{\lambda}} e^{-\sqrt{\frac{\lambda}{3}}\, t}, \tag{9.18}$$

which is finite. Hence there are galaxies to which we can never send a 'hello'. Worse than that, this radius decreases with time. Hence as time passes we are going to be more and more locked down inside our own Galaxy.

Today it seems that this could be the geometry of the universe in the distant future. But we are far from certain about this: we have been repeatedly changing opinions about the cosmological future of the universe in the last few decades, and nothing assures us that we now have the final answer.

- **Cosmological history**

What the data indicate with a certain degree of confidence is that the universe has undergone a radiation-dominated phase, then a matter-dominated phase, and it appears to be moving towards a de Sitter phase. Before these phases things are more uncertain as well.

---

[3] W. de Sitter (1917). 'On the relativity of inertia: Remarks concerning Einstein's latest hypothesis', *Proc. Kon. Ned. Acad. Wet.*, 19: 1217–1225.

Here is what we know of the basic cosmological timeline:

Dominant effect driving dynamics $a(t)$	Quantum gravity ?	Inflation? e^{Ht}	Radiation $t^{\frac{1}{2}}$		Matter $t^{\frac{3}{2}}$		Cosmological constant $e^{\sqrt{\frac{\Lambda}{3}}}$	?
Transitions	?	10^{-32} seconds?	10–10^3 seconds	47 k years	380 k years	1 G year	9.8 G years	13.8 G years
Major events		Big Bang or Big Bounce	Nuclei form		Atoms and CMB form	Galaxies form		Now

There seems to be some indirect evidence that the universe underwent a de Sitter phase also in its early stages, although not everybody is convinced of this. This phase is called 'inflation' and is hypothetically generated by a very hypothetical scalar field called the 'inflaton' (or a set of such fields), which has remained for a while at a value with a high potential energy acting like a transient cosmological constant.

At an even earlier stage, classical general relativity fails because quantum effects become dominant, as Lemaître suspected. There are two main ideas about what could have happened in this quantum phase: a true quantum birth of the universe, or a quantum bounce from a previous contracting phase. I shall briefly touch upon both ideas in the final chapter.

10 The Field of a Mass

10.1 SCHWARZSCHILD METRIC

The metric (7.7) is an approximate solution to the Einstein equations. Due to the complication of these equations, Einstein did not expect that a corresponding exact solution could be found. He was very surprised when, a few weeks after he had published his equations, he received a letter from a young official in the German army with an *exact* solution of his equations (with $\lambda = 0$). Karl Schwarzschild had found the exact solution

$$ds^2 = -\left(1 - \frac{2GM}{c^2 r}\right) c^2 dt^2 + \frac{1}{1 - \frac{2GM}{c^2 r}} dr^2 + r^2 d\theta^2 + r^2 \sin^2 \theta d\phi^2.$$

(10.1)

This metric solves the Einstein equation exactly when $\lambda = 0$, as can be verified explicitly by a laborious but straightforward calculation *[do it!]*.

This metric is a better description of the gravitational field around a static, spherical, massive body than the Newtonian approximation that we studied in Chapter 7. It yields several relativistic effects, which gave the first evidence of the correctness of the theory in the early twentieth century. Furthermore, it opened the door to our understanding of black holes, a topic that I discuss in the next chapter.

Using the notation $d\Omega^2 \equiv d\theta^2 + \sin^2 \theta d\phi^2$ for the metric of a unit sphere and units where $G = c = 1$, the Schwarzschild metric reads

$$ds^2 = -\left(1 - \frac{2M}{r}\right) dt^2 + \frac{dr^2}{1 - \frac{2M}{r}} + r^2 d\Omega^2.$$

(10.2)

Let us examine the physical and geometrical content of this metric. The difference between this and the Minkowski metric (3.110) is twofold.

- **Masses slow down time**

First, the metric component g_{oo} is modified by a correction that becomes stronger as we approach the centre, where the mass is located. We have already seen what this means: clocks slow down as we approach a mass. That is, masses slow down time in their vicinity. Remarkably, this slowing down of clocks has the effect that masses (that follow geodesics in the curved spacetime) fall towards one another.

- **Masses stretch space radially**

The second difference regards the component g_{rr} of the metric. This is a space component, and therefore its alteration represents a change in the spatial geometry. Let us study this in more detail. g_{rr} determines the length of radial lines. From $g_{rr} = 1/(1 - 2M/r) \sim 1 + 2M/r > 1$, we see that the length of radial lines *increases* with respect to Euclidean space as we approach the centre.

What is the resulting geometry? A simple 2d model gives a good intuition about the resulting geometry. Consider the 2d geometry of a funnel like the one shown in Figure 10.1. A moment of reflection shows that the difference between the intrinsic geometry of the funnel and the intrinsic geometry of a plane is precisely the fact that the radial distance increases with respect to the flat one as we approach the centre. The distance between the circle of radius r_1 and the circle of radius r_2 is larger than $r_2 - r_1$, which is the distance between two such concentric circles on a plane.

The spatial geometry around a mass is a three-dimensional analogue of the funnel: the geometrical distance between spheres of coordinate radius r and $r + dr$, which have geometrical areas $4\pi r^2$ and $4\pi(r + dr)^2$, is not dr but rather

$$ds = \sqrt{\frac{1}{1 - \frac{2GM}{r}}} \, dr > dr. \tag{10.3}$$

FIGURE 10.1 A funnel.
Compared with a 2d flat space,
radial distances are expanded,
and more so when approaching
the centre.

Therefore, the distance between the sphere of radius r_1 and the sphere of radius r_2 is

$$D = \int_{r_1}^{r_2} \sqrt{\frac{1}{1 - \frac{2GM}{r}}} \, dr > r_2 - r_1, \qquad (10.4)$$

which is larger than the distance between these two concentric spheres in the Euclidean three-dimensional space. The space around a mass is like a 3d funnel.

10.2   THE KEPLER PROBLEM

To study the relativistic effects of gravity, let us consider the motion of a particle of mass m in the gravitational field of a mass $m \ll M$.

This motion does not depend on m, because particles follow geodesics irrespective of their masses; so, let us put for simplicity $m = 1$.

To start with, we review the same problem in Newtonian physics, using a technique that we can then apply in the relativistic case.

- **The Kepler problem in Newton's gravity**

Since the Newtonian potential is spherically symmetric, angular momentum is conserved. This implies that the particle remains in the plane defined by its radius and its velocity. Without loss of generality, we can adopt the polar coordinates, assuming that this plane is $\theta = \pi/2$. The conserved angular momentum is then given by the quantity (recall we have $m = 1$)

$$L = r v_{tangential} = r^2 \dot{\phi}. \qquad (10.5)$$

Total energy is also conserved and is given by

$$E = \frac{1}{2}v^2 - \frac{GM}{r}. \tag{10.6}$$

Using $v^2 = \dot{r}^2 + (r\dot{\phi})^2$ and the conservation of the angular momentum, this reads

$$E = \frac{1}{2}\dot{r}^2 + \frac{L^2}{2r^2} - \frac{GM}{r}. \tag{10.7}$$

This shows that the radial motion is like the motion of a particle in one dimension, r, in an effective potential

$$V = \frac{L^2}{2r^2} - \frac{GM}{r}. \tag{10.8}$$

The second term is the gravitational potential that determines the gravitational force. The first term is an effective potential (dependent on the angular momentum) that determines the centrifugal force. V has a minimum $(dV/dr|_{r=r_*} = 0)$ at

$$r_* = \frac{L^2}{GM}. \tag{10.9}$$

At this radius there is an orbit where the radius remains constant. This is the circular orbit. The change in the angle with time is directly given by integrating (10.5):

$$\phi(t) = \frac{L}{r_*^2}t = \frac{G^2M^2}{L^3}t. \tag{10.10}$$

Therefore, the angular velocity is

$$\omega_\phi = \frac{G^2M^2}{L^3}. \tag{10.11}$$

The orbits of the planets in the solar system are not circular but are close to circular. We can study them as perturbations of the circular orbits. The radius is not anymore constant but remains close to r_*. We can approximate its dynamics by expanding the potential around the minimum and keeping only the quadratic term. Around r_* we have $V(r) = V_{min} + \frac{1}{2}\omega^2(r - r_*)^2$ where

$$\omega^2 = \frac{d^2V}{dr^2}\bigg|_{r=r_*} = \frac{G^4M^4}{L^6}. \tag{10.12}$$

The motion of the radius is therefore a harmonic oscillation with angular velocity

$$\omega_r = \frac{G^2M^2}{L^3}. \tag{10.13}$$

Notice that this is the same frequency as (10.11):

$$\omega_r = \omega_\phi. \tag{10.14}$$

Therefore, during a full orbit the radius oscillates exactly once. Consequently, the orbit closes, and the perihelion remains constant at the same angular position. These closed orbits are, of course, Keplerian ellipses.

Let us now study the same problem in general relativity.

• *The Kepler problem in Einstein's gravity*

To find the orbits of a body in the gravitational field of a spherical mass, we can integrate the geodesic equation with the Schwarzschild metric. But there is a simpler way, using the integrals of motion as we did in the Newtonian case. We describe the motion of the particle by giving its trajectory in parametrised form, as $x^a(\tau) = (t(\tau), r(\tau), \theta(\tau), \phi(\tau))$. The 4-velocity is then $\dot{x}^a = (\dot{t}, \dot{r}, \dot{\theta}, \dot{\phi})$. The action of a massive particle moving in the Schwarzschild metric is given by (5.4): it is the proper time along its trajectory, that is,

$$S = \int ds = \int \sqrt{g_{ab}\dot{x}^a\dot{x}^b}\ d\tau \tag{10.15}$$

$$= \int \sqrt{-\left(1 - \frac{2GM}{rc^2}\right)c^2\dot{t}^2 + \frac{1}{1 - \frac{2GM}{rc^2}}\dot{r}^2 + r^2\dot{\theta}^2 + r^2\sin^2\theta\,\dot{\phi}^2}\,d\tau, \tag{10.16}$$

where I have kept G and c explicit to keep track of the size of the different terms. As in the Newtonian case, the physics is spherically symmetric; therefore, we can assume without loss of generality that the motion is in the plane $\theta = \pi/2$, which gives

$$S = \int \sqrt{-\left(1 - \frac{2GM}{rc^2}\right)c^2\dot{t}^2 + \frac{1}{1 - \frac{2GM}{rc^2}}\dot{r}^2 + r^2\dot{\phi}^2}\ d\tau \equiv \int \mathcal{L}\,d\tau, \tag{10.17}$$

where $\mathcal{L}$ is a Lagrangian. Since the action does not depend on ϕ, we have the conserved angular momentum

$$L = -\frac{\partial \mathcal{L}}{\partial \dot{\phi}} = -\frac{r^2\dot{\phi}}{\mathcal{L}}. \tag{10.18}$$

Similarly, the Lagrangian does not depend on t, and therefore we have the conserved quantity

$$E = \frac{\partial \mathcal{L}}{\partial t} = -\frac{1 - \frac{2GM}{rc^2}}{\mathcal{L}} c^2 \dot{t}. \qquad (10.19)$$

We can always choose the parametrisation of the orbit $d\tau^2 = -ds^2$, which makes

$$\mathcal{L} = -1, \qquad (10.20)$$

which gives

$$L = r^2 \dot{\phi}. \qquad (10.21)$$

That is, we have the same angular momentum conservation as in the Newtonian case, and therefore the same angular rotation frequency

$$\omega_\phi \equiv \dot{\phi} = \frac{L}{r^2} \qquad (10.22)$$

and

$$E = -\left(c^2 - \frac{2GM}{r} \right) \dot{t}. \qquad (10.23)$$

Using these two equations, we replace $\dot{t}$ and $\dot{\phi}$ in (10.20) and we obtain

$$-\frac{E^2}{1 - \frac{2GM}{rc^2}} + \frac{\dot{r}^2}{1 - \frac{2GM}{rc^2}} + \frac{L^2}{r^2} = -c^2, \qquad (10.24)$$

which with minimal algebra gives

$$\frac{1}{2}\dot{r}^2 - \frac{GM}{r} + \frac{L^2}{2r^2} - \frac{GML^2}{c^2 r^3} - \frac{E^2 + c^2}{2} = 0. \qquad (10.25)$$

This equation shows that the motion of r is like the motion of a particle in an effective potential

$$V = -\frac{GM}{r} + \frac{L^2}{2r^2} - \frac{GML^2}{c^2 r^3}, \qquad (10.26)$$

plus the irrelevant constant term $V_o = -(E^2 + c^2)/2$. See Figure 10.2. Notice that this effective potential is like the Newtonian effective

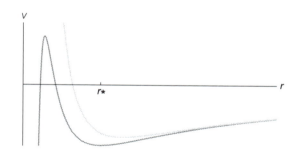

FIGURE 10.2 The effective potential for a massive particle around a central mass. The curve further to the right is the Newtonian one.

potential (10.8) with one term added. This term corresponds to an attractive force

$$F = -\frac{3GML^2}{c^2r^4}.$$ (10.27)

The relativistic effect of gravity on an object orbiting a central mass M is simply this additional attractive force.

- **Effects of the relativistic correction to the force**

Let us study the characteristics of this relativistic gravitational force. First, it is proportional to L^2, namely to the radial velocity. This means that it is a magnetic-like force: it is not felt by a mass without angular velocity. Second, it is inversely proportional to c^2; therefore, it is a relativistic effect and is small for non-relativistic velocities. Third, it is proportional to r^{-4}, which means that it becomes important – in fact, dominant – at small radii. In the solar system, the planet with the smallest radius and the largest angular velocity is Mercury; therefore, we may expect that Mercury is the first planet where the effect of this relativistic force has a chance to be detected.

Let us see what is the effect of this force on the orbit. Repeating what we did in the Newtonian case, we find that there is a circular orbit at the local minimum of the potential, namely when at radius r_* satisfying

$$\left.\frac{dV}{dr}\right|_{r=r_*} dr = \frac{GM}{r_*^2} - \frac{L^2}{r_*^3} + \frac{3GML^2}{c^2r_*^4} = 0.$$ (10.28)

The second derivative of the potential at the point r_* is

$$\omega_r^2 = \frac{d^2 V}{dr} \bigg|_{r=r_*} dr = -2\frac{GM}{r_*^3} + \frac{3L^2}{r_*^4} - \frac{12GML^2}{c^2 r_*^5}, \tag{10.29}$$

which using the equality (10.28) gives

$$\omega_r^2 = \frac{L^2}{r_*^4} - \frac{6GML^2}{c^2 r_*^4} = \omega_\phi^2 \left(1 - \frac{6GM}{c^2 r_*}\right). \tag{10.30}$$

Therefore, a relativistic effect breaks the Newtonian equality between ω_ϕ and ω_r. During a full oscillation of the radius, namely in the time $T_r = 2\pi/\omega_r$, the angle changes by an amount

$$\phi = \omega_\phi / T_r = \frac{2\pi}{1 - \frac{6GM}{c^2 r_*}} \sim 2\pi + \frac{6\pi GM}{c^2 r_*} \tag{10.31}$$

(to first order in c^{-2}). Therefore, at every orbit the perihelion advances by the angle (again to first order in c^{-2})

$$\delta_\phi \sim \frac{6\pi GM}{c^2 r_*}. \tag{10.32}$$

• **The first test of general relativity**

For Mercury, the radius of the orbit is $r \sim 55 \times 10^6$ km and the mass of the Sun gives $GM/c^2 r \sim 1.45$ km, which gives $\delta_\phi \sim 0.104$. Mercury makes 415 revolutions per century, which gives a precession of the perihelion of the angle

$$\Delta \phi \sim 43'' \text{per century.} \tag{10.33}$$

This is in excellent agreement with the *measured* precession unaccounted for by Newtonian theory.

$$\Delta \phi_{measured} \sim (42'' \pm 1'') \text{per century.} \tag{10.34}$$

This precession had already been measured before Einstein completed his theory. Finding that the theory accounted precisely for the unaccounted part of the measured value of the precession was the first triumph of general relativity and gave Einstein strong confidence in the theory he was developing.

In his long search for field equations, Einstein recalculated this precession several times at each tentative field equation. When he found the field equations leading to the correct value (not using the Schwarzschild metric, which he did not yet have, but the approximate solution $ds^2 = -(1 - 2M/r)dt^2 + (1 + 2M/r)dr^2 + r^2 d\Omega^2$), he convinced himself that they were the right ones.

10.3 DEFLECTION OF LIGHT BY THE SUN

The first (of many) spectacular predictions of *new* phenomena by general relativity was the deflection of light by the Sun.

The easiest way to understand this deflection and compute it is to use Fermat's principle: light rays follow trajectories that minimise the travel time between the source and the arrival point. (The reason is simple: light is a wave, and these are the trajectories along which interference is constructive.)

If light emitted by a star passes near the Sun, it probes the spacetime geometry where it is altered by the Sun's mass. As we have seen by analysing the Schwarzschild metric, in these coordinates a mass produces two geometrical effects in its vicinity: the slowing down of time and the curving of the spatial geometry which stretches the radial direction. Both have the effect of slowing down light that passes near the Sun, in comparison with light that passes at a larger distance from it. Therefore, to minimise travel time, a light ray should keep itself at a distance from the star. But not too much, because its path would become too long otherwise. Let's put this into equations.

In general coordinates the speed of light is not c, but is rather given by the equation $ds = 0$. Since most of the trajectory of light that passes near the Sun is nearly radial, let's take the approximation $d\Omega = 0$. This gives

$$ds^2 = -\left(1 - \frac{2GM}{rc^2}\right)c^2 dt^2 + \frac{1}{1 - \frac{2GM}{rc^2}}dr^2 = 0, \qquad (10.35)$$

which gives

$$v(r) = dr/dt = c\left(1 - \frac{2GM}{rc^2}\right). \qquad (10.36)$$

This is the speed of light near the Sun in these coordinates: light slows down near the star.

Let us now simplify the trajectory of the ray by making it go straight (in these coordinates) until the closest point to the Sun, take a sharp angle α, and come out straight again, as in Figure 10.3. The time light takes to travel from the star (which is very distant) to the deflection point at a distance b from the centre of the Sun is

$$T = \int_\infty^0 \frac{ds}{v(r)} = \int_\infty^0 \frac{ds}{c\left(1 - \frac{2GM}{rc}\right)} = \int_\infty^0 \frac{ds}{c\left(1 - \frac{2GM}{\sqrt{b^2+s^2}c^2}\right)}.$$

(10.37)

Expanding for small c^{-2} gives

$$T = \int_\infty^0 \frac{ds}{c}\left(1 + \frac{2GM}{\sqrt{b^2 + s^2}c^2}\right).$$

(10.38)

The variation of this time with b is

$$\left.\frac{dT}{db}\right|_{velocity} = \frac{2GM}{c^3}\int_\infty^0 ds\frac{2b}{(b^2 + s^2)^{\frac{3}{2}}}$$

$$= \frac{4GM}{bc^3}\int_\infty^0 dx\frac{1}{(1 + x^2)^{\frac{3}{2}}} = -\frac{4GM}{bc^3}.$$

(10.39)

(The last step is from $(d/dx)\, x/\sqrt{1 + x^2} = (1 + x^2)^{-3/2}$.) By increasing b, the relativistic effect decreases the travel time by this amount.

On the other hand, increasing b increases the travel distance. This can be simply estimated (see Figure 10.3) by $L \sim b\sin\alpha \sim b\alpha$;

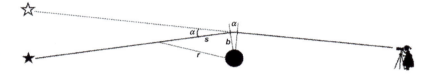

FIGURE 10.3 Deflection of light by the Sun. The white star is the apparent position of the star in the sky.

this is crossed in a time $T \sim b\alpha/c$, and, therefore, the variation of the transit time due to the change in length is

$$\frac{dT}{db}\bigg|_{length} = \frac{\alpha}{c}. \tag{10.40}$$

By Fermat's principle, the ray that light follows is where the total variation vanishes, that is,

$$\frac{dT}{db} = \frac{dT}{db}\bigg|_{velocity} + \frac{dT}{db}\bigg|_{length} = -\frac{4GM}{bc^3} + \frac{\alpha}{c} = 0, \tag{10.41}$$

which gives

$$\alpha = \frac{4GM}{c^2 b}. \tag{10.42}$$

For a star whose trajectory passes very close to the surface of the Sun, b is the radius of the Sun, which is 7×10^5 km, while $GM/c^2 \sim 1.45$ km, giving

$$\alpha \sim 1.75''. \tag{10.43}$$

This was Einstein's prediction. In 1919, an expedition led by Arthur Eddington to measure this deflection during an eclipse (without eclipses, the stars near the Sun are not visible in the clear day sky) confirmed the reality of this deflection, a feat that elevated him to instant global fame.

Exercise: Einstein actually had been lucky. A few years before, he had predicted a deflection half the value (10.42), because he used the metric (7.7) instead of the right one. An expedition to test this (wrong) prediction could not measure anything because of clouds during the eclipse. This lucky mishap gave him the time to correct the prediction before the testing. [*Show why the metric (7.7) gives only half the effect.*]

10.4 NEAR HORIZON ORBITS

For a given mass M, the radius

$$r_S = \frac{2GM}{c^2} \tag{10.44}$$

is called the Schwarzschild radius. When the radius of the orbit is large compared to the Schwarzschild radius, the only relativistic effects are the small corrections to the Kepler orbits studied in the previous section. Closer to the Schwarzschild radius, interesting things happen.

At shorter distances from the Schwarzschild radius, the relativistic effects are much stronger. To study them, let us reconsider the effective potential (10.26). The plot of this potential, shown in Figure 10.2, shows that there is a second extremum, absent in the Newtonian case: with it, the force becomes again attractive, overcoming the centrifugal force due to the angular momentum. The two extrema are at the radii

$$r_\pm = \frac{c^2 L^2 \pm \sqrt{c^4 L^4 - 12c^2 G^2 L^2 M^2}}{2c^2 GM}. \tag{10.45}$$

For large enough angular momentum L, the square root is positive and the potential has two extrema, as shown in Figure 10.2. The larger extremum (at r_+) is a minimum and determines the stable circular Keplerian orbit considered above. But there is a second extremum (at r_-), which corresponds to a maximum. This is a lower *unstable* circular orbit.

- **Minimal stable orbit**

The radius of the stable Keplerian orbit decreases with L^2; but if L^2 is too small, the argument of the square root becomes imaginary and the potential has no minima. (Figure 10.2 is not qualitatively correct anymore.) This happens when $L^2 = 12G^2M^2/c^2$, and at this value of L^2 the radius of the minimum is equal to the radius of the maximum and has the value

$$r = \frac{6GM}{c^2} = 3r_S. \tag{10.46}$$

This has an important consequence: there are no stable orbits below the radius $\frac{6GM}{c^2}$, which is three times the Schwarzschild radius. When spiralling matter reaches this radius, any further minimal loss of energy has the consequence that matter plunges into the hole, as

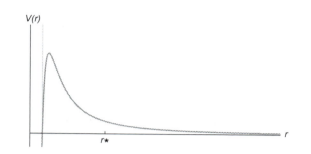

FIGURE 10.4 The effective potential for a light ray around a black hole

it has no stable orbits anymore on which to sit. The existence of this minimal stable orbit in the accretion discs of astrophysical black holes has been observed and is one of the clear signs that indicates that objects we see in the sky are indeed well described by this mathematics.

• *Light's orbits*

Before concluding this section, let us study also the trajectory of light. We can repeat the same analysis as above, with the only difference that, instead of $ds^2 = -1$, we have $ds^2 = 0$, which is the light ray equation (4.8). This yields a slightly different effective potential:

$$V = \frac{L^2}{2r^2} - \frac{GML^2}{c^2r^3}.$$

(10.47)

Compared with the effective potential for massive particles, the effective potential for light lacks only the Newtonian term. See Figure 10.4.

It is sometimes stated that in the non-relativistic limit light is attracted by matter. This is not correct: in the non-relativistic limit, the relativistic term $\frac{GML^2}{c^2r^3}$ goes to zero, and only the centrifugal term remains. This gives straight light rays.

The effective potential now has a maximum (see Figure 10.4) at

$$r = \frac{3GM}{c^2} = 1.5\, r_S.$$

(10.48)

This means that light can orbit around a mass at one and half its Schwarzschild radius. Light rays are very much distorted by the strong attraction of a mass, in the region just outside the Schwarzschild radius.

In the next chapter we get to r_S: we study what happens at this radius and inside it, namely the physics of black holes. Before concluding this chapter, however, I add two brief sections on the effect of a mass *at very large distances*, where the cosmological term of the Einstein equations becomes relevant, and on the field of a *charged and rotating* mass.

10.5 COSMOLOGICAL FORCE

An exact solution of the Einstein equations *with* cosmological constant is

$$ds^2 = -\left(1 - \frac{2M}{r} - \frac{\lambda}{3}r^2\right) dt^2 + \frac{1}{1 - \frac{2M}{r} - \frac{\lambda}{3}r^2} dr^2 + r^2 d\Omega^2.$$

$$(10.49)$$

The difference with the Schwarzschild case is that the Newtonian potential $-\frac{M}{r}$ is modified by the addition of the term $-\frac{\lambda}{6}r^2$. This determines a *repulsive* force per unit test mass

$$F_\lambda = \frac{\lambda}{3} r.$$

$$(10.50)$$

This force is weak because λ is small, but it increases with distance. Therefore, it manifests itself only at large – in fact, cosmological – distances, hence the name of the constant.

Thus, gravity is attractive at small distances but becomes repulsive at large distances. The (unstable) equilibrium point from a body of mass M is at

$$\frac{GM}{r^2} = \frac{\lambda}{3} r,$$

$$(10.51)$$

that is,

$$r = \sqrt[3]{\frac{3GM}{\lambda}}. \tag{10.52}$$

This force (not any mysterious 'quantum vacuum energy' or 'dark energy') is the reason why the expansion of the universe is currently accelerating. (See the comment at the end of Section 4.3.)

10.6 KERR–NEWMAN METRIC AND FRAME DRAGGING

The Schwarzschild metric describes the spacetime around a non-rotating mass. It took a long time to discover the spacetime metric around a rotating mass. Here it is. For completeness, I give the metric of a rotating mass M with angular momentum $J = ac^2M$ and electric charge Q. This is:

$$ds^2 = \rho^2 \left(\frac{dr^2}{\Delta} + d\theta^2 \right) - \frac{\Delta}{\rho^2}(dt - a \sin^2\theta \, d\phi)^2 \tag{10.53}$$
$$+ \frac{\sin^2\theta}{\rho^2}((r^2 + a^2)d\phi - a \, dt)^2,$$

where

$$\rho^2 = r^2 + a^2 \cos^2\theta, \quad \Delta = r^2 - 2GMr + a^2 + Q^2G. \tag{10.54}$$

This metric is called the Kerr–Newman metric, after the New Zealand mathematician Roy Kerr and the American relativist Ted Newman. It solves the coupled Einstein–Maxwell equations (with $\lambda = 0$). Notice that this metric is not diagonal. It has a non-vanishing $g_{t\phi}$ term.

● **Frame dragging**

To illustrate an effect of a rotating body, consider this metric at the North Pole of the Earth. The total charge of the Earth is negligible $(Q = 0)$, and at the North Pole we can write $\cos\theta = 1$. Let us fix the value of the radius to be the Earth radius and consider the resulting 3d metric in t, θ, ϕ coordinates, keeping only the second order in θ and using $R = \rho\theta$. We obtain

$$ds^2 = \left(1 - \frac{2GM}{\rho^2} + R^2\frac{a^2}{\rho^4}\right)dt^2 - dR^2 - R^2 d\phi^2 + 2R^2\frac{2GMa}{\rho^4}dtd\phi,$$

$$(10.55)$$

where ρ is a constant, close to the radius of the Earth. The parentheses in the dt^2 terms are a small modification of the standard gravitational time dilatation. The second and the third terms give the standard metric of the plane in polar coordinates. The interesting term is the last one. What does it represent? Just look back at equation (3.122) and the discussion surrounding it, and its meaning becomes transparent: it indicates that the frame defined by these coordinates is rotating with respect to an inertial frame. The angular velocity is

$$\omega = \frac{2GM}{c^2\rho^4}a.$$

$$(10.56)$$

Since the metric is stationary (time independent) in the coordinates t, r, θ, ϕ, the inertial frame rotates with respect to the stationary frame. The fixed stars are stationary in the stationary frame. Therefore, an inertial frame at the North Pole rotates with respect to the fixed stars!

The Earth is a sphere rotating with the angular velocity $\omega_\oplus = 1/\text{day}$. Its angular momentum is of the order $J_\oplus \sim r_\oplus^2 M_\oplus \omega_\oplus$, hence $a_\oplus = J/(M_\oplus c^2) = r^2\omega_e/c^2$ and a reference frame at the North Pole rotates with an angular velocity

$$\omega \sim \frac{r_s}{r_\oplus}\omega_\oplus \sim \frac{1.5 \text{ cm}}{6,000 \text{ km}}1/\text{day} \sim 2 \times 10^{-8}/\text{day}.$$

$$(10.57)$$

That is, a frame at the North Pole fixed with respect to the fixed stars is actually rotating with respect to an inertial frame at this (very small) angular velocity.

Newton's bucket, III. Remember Newton's bucket argument: the concavity of the water in a rotating bucket detects absolute rotation. Hence the existence of absolute space. The above result shows that the water is concave when it is rotating with respect to the local value of the gravitational field, which in turn is affected by the big rotating mass of the Earth: *Newton's space is the gravitational field.*

11 Black Holes

Something strange happens to the Schwarzschild metric at the radius

$$r_S = \frac{2GM}{c^2} = 2m. \qquad (11.1)$$

In this section, I use $G = c = 1$ and I indicate the mass of the black hole with a lower-case m. The Schwarzschild radius is therefore just $r_S = 2m$.

At this radius the component g_{oo} of the metric vanishes, and the component g_{rr} becomes infinite. That is, clocks stop, and the distance between the sphere of area $4\pi r^2$ and the sphere at infinitesimal coordinate distance dr from it is infinite. Einstein thought that the world ended there: no spacetime with $r < 2m$.

He was wrong.

Still in the 1970s, the Nobel Prize physicist Steven Weinberg wrote in his book on general relativity that the question may well be just academic, because the Schwarzschild solution holds *outside* a spherical mass. The Earth's Schwarzschild radius is $r_S \sim 1$ cm, hence to have a regime where this question is relevant we should reach a density comparable to concentrating the entire mass of the Earth in about 1 cm^3! It seemed unlikely that anything like that could exist in the universe.

Today we know better. An extensive discussion during the 1960s clarified the phenomena that the theory predicts around $r = r_S$, leading to the understanding of the geometry and physics of the horizons. At first these peculiarities of the Schwarzschild metric were suspected to be artefacts of its symmetry. Roger Penrose developed mathematical methods and proved a theorem[1] that ended

---

[1] R. Penrose, 'Gravitational collapse and space-time singularities', *Physical Review Letters*, vol. 14, no. 3 (1965), pp. 57–59.

up convincing the research community that they were in fact independent from the symmetry and generic, work recognised by the 2020 Nobel Prize.

In the last few decades a great number of objects have effectively been found in the sky, where these phenomena indeed happen: we call them *black holes*. Black holes abound in the universe. The ones observed so far have masses that range from a few solar masses to a billion solar masses, but black holes of other sizes may well exist, for instance, very small black holes produced in the primordial universe.

Evidence for the existence of objects in the sky that are well described by the mathematics of black holes is abundant: it comes from powerful X-rays emitted by the 'accretion discs' formed by matter spiralling and falling into the hole, by detection of the gravitational waves produced by black hole mergers, and even by direct radio telescope imaging of the immediate surroundings of a large black hole.

One of the most striking elements of evidence for these objects comes from observation of the stars orbiting the central black hole of our Galaxy. A simple Newtonian analysis of their Keplerian orbits indicates that the object around which they orbit has a mass of the order of 4 million solar masses, concentrated in 125 AU (Astronomical Units: the Sun–Earth distance), in a location in good correspondence with the powerful compact radio source Srg A*, which itself has an apparent size of less than one AU and no detectable proper motion. This indicates that the mass of 4 million suns is concentrated in a region smaller than the Earth orbit. It is hard to imagine anything other than a black hole could describe such an object. Reinhard Genzel and Andrea Ghez have received the 2020 Nobel Prize for these observations.

11.1 AT THE HORIZON

Reconsider the example at the end of Section 3.1.3, illustrated in Figure 3.4. The example illustrates what happens with the

Schwarzschild coordinates. In the actual physical spacetime there is a surface $r = r_S$, but the Schwarzschild coordinates avoid it, precisely like the coordinates y, x of the example avoid the line $X = 0$. More precisely, this surface is reached only when $t \to \infty$.

Conversely, the entire line $(r = r_S, t, \theta, \phi)$ of the Schwarzschild metric, at fixed values of θ and ϕ, *corresponds to a single point in physical spacetime*, just as all points with polar coordinates $(0, \phi)$, for any ϕ, represent the same North Pole.

To see that this is the case, we must pass to better coordinates, similar to the 'good' Cartesian coordinates X, Y of the example of Section 3.1.3.

- **Painlevé–Gullstrand coordinates**
A set of better coordinates is obtained simply by changing the time coordinate. Define

$$t_* = t + 2\sqrt{2mr} + 2m \ln \frac{\sqrt{r/2m} - 1}{\sqrt{r/2m} + 1}. \tag{11.2}$$

These are called the Painlevé–Gullstrand coordinates. See Figure 11.1. They allow us to describe the region of spacetime at and past $r = 2m$. Easily

$$dt = dt_* - \frac{\sqrt{2m/r}}{1 - 2m/r} dr. \tag{11.3}$$

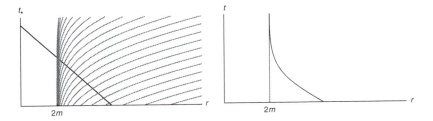

FIGURE 11.1 Left: the lines of constant t coordinate seen in the coordinates r, t_*. The dark line represents an infalling trajectory. Notice that it crosses $r = r_S$ only at $t = \infty$. Right: the same trajectory in the plane of the Schwarzschild coordinates r, t. Notice that in these coordinates the trajectory never reaches $r = 2m$.

Replacing in the Schwarzschild metric yields the nice form

$$ds^2 = -dt_*^2 + (dr + \sqrt{2m/r}\,dt_*)^2 + d\Omega^2. \tag{11.4}$$

These are better coordinates (not yet the best ones, as we shall see later on) to understand what happens at $r = 2m$. Let us study this region.

- **Light cones at the Schwarzschild radius**

To this aim let us study the radial $(d\Omega = 0)$ null $(ds = 0)$ trajectories of light. These are given by

$$dt_*^2 = (dr + \sqrt{2m/r}\,dt_*)^2. \tag{11.5}$$

There are two solutions to this equation, corresponding to the ingoing and the outgoing light rays, respectively:

$$\frac{dr}{dt^*} = \pm 1 - \sqrt{2m/r}. \tag{11.6}$$

For very large r, we have

$$\frac{dr}{dt_*} \sim \pm 1. \tag{11.7}$$

This is the standard Minkowski result: light travels along outgoing rays $r = t$ and ingoing rays $r = -t$. Thus, the $+$ sign gives the outgoing rays, the $-$ sign gives the ingoing ones. As the radius r decreases, $\frac{dr}{dt_*}$ decreases. For the ingoing rays, the sign of $\frac{dr}{dt_*}$ remains always negative. This is what is expected: ingoing rays have decreasing radius. But for the outgoing rays, the sign of $\frac{dr}{dt_*}$ changes at $r = 2m$. What does this mean?

It means that for $r < 2m$ the 'outgoing' rays are *infalling* in the sense that they move towards smaller radius.

That is, suppose you light up for one instant a sphere of constant radius r. There are two light fronts that leave the sphere and move away from it: one outside and one inside. These are represented by two null rays in the (r, t_*) plane. If r is larger than $2m$, one of the rays moves towards larger r and one moves towards smaller r. But if r is smaller than $2m$, both move towards a smaller radius! (If the

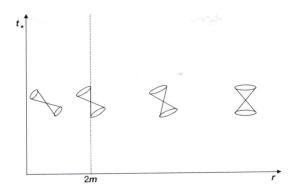

FIGURE 11.2 Light cones in the t_*, r coordinates

sphere was made of matter, it would move towards a smaller radius as well, remaining between the two rays.) Gravity is so strong that light cannot escape: gravity forces it to fall in, no matter what.

This is summarised in Figure 11.2. The figure neatly illustrates many features of the extended Schwarzschild solution by showing the shape of the light cones as we move across the horizon. Since matter can only move in timelike directions, we see from the picture that inside $r = 2m$ it is impossible to remain at fixed radius: any matter is forced to fall in. Gravity is so strong that remaining at fixed radius is impossible. Even less possible is to move out. Light itself cannot avoid falling towards smaller radius.

- **Trapping horizons**

Inside the black hole, the area of any light front emitted by an $r = constant, t = constant$ 2-sphere shrinks. In general, a region where this is true for any two-sphere is called 'trapped'. Penrose has introduced and precisely defined this idea of trapped region and has proven that once a trapped region forms, it is impossible, according to general relativity and if energy density is positive, to avoid the collapse towards a singularity.

The boundary of the trapped region of the Schwarzschild metric is the 3d surface $r = 2m$. It is a null surface. Light can move along it. A light front can actually stay on $r = 2m$, keeping the same area.

This three-dimensional surface $r = 2m$ is the boundary of the black hole. It is called the 'horizon' of the black hole. The reason for this name is that as long as things remain stationary in this manner, an external observer cannot see things that are beyond the horizon – just as we cannot see beyond the horizon on Earth. Light emitted from things inside the black hole falls in and does not get out.

Obviously, this does not mean that there is nothing beyond the horizon: it only means that to see what is there, we have to go there (cross the horizon).

• **Event horizons**

The $r = 2m$ surface has another interesting property, besides being a trapping horizon. It is the inner boundary of the part of spacetime from which light rays can escape to infinity. Technically, this is expressed as follows: it is *the boundary of the past of future (null) infinity*. (Future null infinity is the ensemble of directions in which light can escape; its past is the region from which light can escape.) In a general spacetime, *the boundary of the past of future null infinity* is called an 'event horizon'. Therefore, the Schwarzschild trapping horizon $r = 2m$ is *also* an event horizon.

For the Schwarzschild solution, the trapping horizon and the event horizon coincide, but this is not true in general. A simple example where they do not coincide is the following. Say at some time matter falls into a black hole, increasing its mass from m to $m + \Delta m$. A point just outside $r = 2m$ shortly before the fall of the matter into the hole can be outside the trapping horizon, because light emitted from it moves towards increasing radius; but it can also be inside the event horizon, because when the black hole becomes bigger this light ray finds itself inside the radius $r = 2(m + \Delta m)$ and therefore never escapes.

There is an important conceptual difference between the notion of 'trapping horizon' and the notion of 'event horizon'. The first depends only on the present (light rays fall towards smaller radius), while the second depends on what happens in the future. The location of an event horizon depends on the full future (or distant) state of the field. On the contrary, to know the location of a trapping horizon is sufficient to look at the region of the black hole.

The Schwarzschild solution is only an approximation: it disregards quantum effects. We do not know what happens to a real black hole in the distant future. It is possible that light rays that fell in might ultimately escape. If this is the case, real black holes do not have event horizons. But the surface $r = 2m$ is still a trapping horizon. Hence event horizons are less relevant than trapping horizons.

• **Eddington–Finkelstein coordinates**

Another simple set of coordinates that cover the entire region outside and inside the black hole can be nicely written using mixed coordinates v, r analogous to the coordinates used in (3.118) for the Minkowski space. The Schwarzschild geometry in these coordinates reads very simply

$$ds^2 = (1 - 2m/r) \, dv^2 - 2dv \, dr + r^2 d\Omega^2. \qquad (11.8)$$

The relation with the Schwarzschild coordinates is given by the change of variable

$$t = v - r - 2m \log |r/2m - 1|. \qquad (11.9)$$

Exercise: Derive (11.8) from (10.2) using this change of variable.

The lines of constant v, θ, ϕ are the radial light rays coming from infinity and falling into the black hole.

11.2 INSIDE THE BLACK HOLE

What is there inside the black hole?

• **_The shape of the interior_**

The interior of a black hole is described by the line element (11.4) with $r < 2m$. We can undo the change of coordinates (11.2) and return to Schwarzschild coordinates. Therefore, for $r < 2m$ the Schwarzschild line element (10.2) describes the interior of the black hole: it is only the boundary between the two regions that goes wrong in these coordinates.

Since the metric components g_{tt} and g_{rr} change sign at $r = 2m$, inside the hole r becomes a temporal variable and t a space variable. This has no particular physical meaning: it is only arbitrary naming of local coordinates.

The symmetries of the metric are different inside and outside. The independence of the metric coefficients on the t variable has a different meaning outside and inside. Outside, there is temporal translation symmetry. That is, spacetime is static. Inside, there is a translational symmetry in a spatial direction. The dependence of the metric on the temporal r variable means that the metric is not anymore time-independent in the interior. The surfaces $r = constant$ are

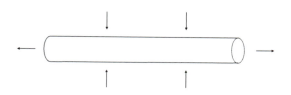

FIGURE 11.3 The geometry of the interior of the black hole: a constant time (constant r) surface is a 3d cylinder formed by a 2-sphere times a line. As time passes (r decreases), the cylinder becomes longer and more narrow.

spacelike surfaces. Their 3d (positive definite) metric is

$$ds^2 = \left(\frac{2m}{r} - 1\right) dt^2 + r^2 d\Omega^2. \tag{11.10}$$

This is the metric of a 3d cylinder: a sphere of radius r times a line with line element $\left(\frac{2GM}{r} - 1\right) dt^2$. As time passes, r decreases. The sphere shrinks, and the length of the cylinder increases.

This is the interior geometry of a black hole: a long cylinder with a radius that decreases in time and a length that increases in time. As time passes (r decreases), the cylinder becomes increasingly longer and narrower. See Figure 11.3. In a sense, the external passing of time 'feeds' the elongation of the internal cylinder across the horizon.

• **Falling in**

What happens entering the black hole?

Crossing the horizon does not involve any difficulty: locally spacetime is always flat, and therefore a point on the horizon is a point that is locally completely normal. What happens next?

The fall is free fall, so a small interstellar ship falling would feel no force. However, the curvature increases as it approaches the centre. (The components of the Schwarzschild metric, such as Newton's potential, grow as $1/r$ when $r \to 0$, their first derivatives as $1/r^2$, and their second derivatives and therefore curvature as $1/r^3$.) The larger the curvature, the smaller the region in which spacetime is approximated by Minkowski. When the radius of curvature becomes of the order of the size of the spaceship, the inertial motion of its different parts no longer respects the geometry of the spaceship: in other words, the spaceship undergoes 'tidal' forces that deform it. Since the attraction towards the centre is stronger at smaller radii, the interior of the spaceship accelerates more than the outside, and the tidal forces radially stretch the spaceship. Since these forces diverge with $r \to 0$, the ship, or any structure, is destroyed before reaching the centre.

How long does this process take?

Consider the case of an object falling radially, with vanishing angular momentum. By inspection, the maximal proper time of a radial infalling trajectory is characterised by $dt = 0$; therefore, the proper time from crossing the horizon to the centre is

$$T = \int ds = \int_{2m}^{0} \sqrt{-g_{rr}(r)}\, dr = \int_{2m}^{0} \frac{dr}{c\sqrt{1 - \frac{2GM}{c^2 r}}} = \frac{\pi Gm}{c^3}.$$

(11.11)

For a stellar black hole, where the Schwarzschild radius is of the order of kilometres, this is of the order of microseconds. Not much time to appreciate the views.

But for the largest black holes detected so far that have Schwarzschild radius 10^9 times a stellar black hole, this falling time can last for hours. Therefore a physicist can in principle enter the horizon and do all sort of measurements and detections in the interior. It is true that she is most likely going to die soon, crushed in the small r region; but we all die anyway, even outside the horizon.

● **Towards the centre**

The order of magnitude of the curvature, as we have seen, is $GM/c^2 r^3$. The curvature reaches a Planckian value, namely $\frac{GM}{c^2 r^3} \sim \frac{c^3}{\hbar G}$, when

$$r \sim \sqrt[3]{\frac{G^2 \hbar M}{c^5}}.$$

(11.12)

When the radius reaches this scale, quantum effects are likely to become relevant and cannot be neglected anymore: classical general relativity is not anymore reliable. Notice that (11.12) indicates that this can happen at a radius much larger than the Planck length. I will touch upon these quantum effects in the final chapter.

One often reads that general relativity predicts a 'singularity' at the centre of the hole; whatever this means, this is not relevant for the natural world, because quantum mechanics becomes relevant and changes the picture before reaching such a hypothetical 'singularity'.

11.3   WHITE HOLES

The Einstein equations are invariant under time inversion $t \mapsto -t$. That is, if we film a solution of the Einstein equation and project the film backward in t, we get a metric that is again a solution of the Einstein equations.

The Schwarzschild metric (10.2) does not change replacing t with $-t$, therefore the exterior of a black hole is invariant under time reversal symmetry. But not so the horizon and the interior of a black hole. The time inversion of the black hole metric (11.4) is the metric

$$ds^2 = -dt_*^2 + (dr - \sqrt{2m/r}\, dt_*)^2 + r^2 d\Omega^2, \tag{11.13}$$

which is a different extension of the exterior than (11.4): while (11.4) extends it to the future, (11.13) extends it to the past. The extension to the past is called a white hole. Its light cone structure is simply the upside version of Figure 11.2: see Figure 11.4. In mixed coordinates u, r, the region formed by the white hole interior and exterior is given by

$$ds^2 = (1 - 2m/r)\, du^2 + 2du\, dr + r^2 d\Omega^2 \tag{11.14}$$

to be compared with (3.119) and (11.8). The lines of constant u, θ, ϕ are the radial light rays coming out from the white hole.

Since the Schwarzschild metric is invariant under $t \mapsto -t$, outside the horizon of a white hole there is precisely the same metric as outside a black hole. The exterior of a white hole behaves exactly

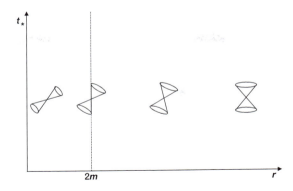

FIGURE 11.4 White hole light cones in the t_*, r coordinates

like that of a black hole: it has a mass, it is attractive, things can orbit around it, and so on. It is only at the horizon itself that things differ.

- **The relative location of the black and white holes with respect to the exterior**

The above is confusing at first. How can things differ arriving at the horizon, if everything is the same outside the horizon? The answer is that the metric (10.2) can be continued towards two distinct regions: one in the future, one in the past.

The key is to remember that an $r = 2m$ line in Schwarzschild coordinates, for fixed θ and ϕ and arbitrary t, is *a single point* in physical spacetime. It is a point where *two* null boundaries of the external Schwarzschild geometry meet. The Schwarzschild time coordinate t goes to minus infinity at the boundary with the white hole, and to plus infinity at the boundary with the black hole. These infinities are just pathologies of the coordinate: the geometry is regular at the boundaries of the external Schwarzschild region. The external Schwarzschild region is thus bounded by two horizons. One is at large positive t, the other at large negative t. See Figure 11.5.

The location of the two extensions of the external Schwarzschild metric becomes clearer by recalling the Rindler coordinates in Minkowski space, illustrated in Figure 3.8. These cover only a wedge of Minkowski space: the Rindler wedge $x > |t|$. This wedge is bounded by

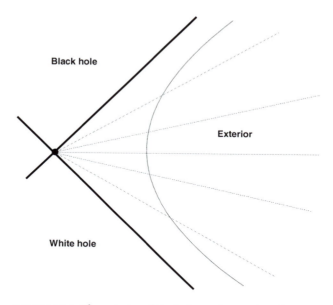

FIGURE 11.5 The exterior of the hole can be continued in two
directions: into a black hole across the $r = 2m, t = +\infty$ surface; and into
a white hole across the $r = 2m, t = -\infty$ surface. The straight lines are
$t = constant$ surfaces. The curved line is a constant-radius surface.
Notice that all spheres $r = 2m$ for any arbitrary finite t actually
represent the same single sphere in spacetime, here indicated by the dot
joining the two horizons (bold lines).

two null lines that meet at the origin. The two null lines are respectively
at $\tau = \pm\infty$. The entire $\rho = 0$ line in Rindler coordinates, for any arbitrary
finite τ, is *a single point* in Minkowski space: the point where the two
null boundaries of the wedge cross. This is exactly the same situation as
for the two boundaries of the Schwarzschild exterior geometry. Compare
Figures 3.8 and 11.5.

- **Schwarzschild and Rindler**

The similarity between Schwarzschild and Rindler spacetime is more
than an analogy. In fact, it turns out that near $r = 2m$ the Schwarzschild
metric in the t, r plane is precisely the Rindler metric! To see this, let us
write

$$r = 2m + x \tag{11.15}$$

and assume that we look at the metric for $x \ll 2m$, namely very near
$r = 2m$. Then we can expand

$$1 - 2m/r = 1 - 2m/(2m + x) = 1 - 1/(1 + x/2m)$$
$$\sim 1 - (1 - x/2m) = x/2m \tag{11.16}$$

so that the Schwarzschild metric (with $d\Omega^2 = 0$) reads

$$ds^2 = -\frac{x}{2m}dt^2 + \frac{2m}{x}dx^2. \tag{11.17}$$

Changing variables $x = \rho^2/(8m)$ and $t = 4m\tau$ gives

$$ds^2 = d\rho^2 - \rho^2 d\tau^2, \tag{11.18}$$

which is precisely the Rindler metric. That is, in the vicinity of the horizon, if we look only at the radial and temporal coordinates, the Schwarzschild metric looks precisely like a Rindler metric.

This is physically comprehensible as follows. Imagine that you remain at a very small distance from the horizon. To stay there without falling, you need rockets, namely you are not in free fall: you need a constant acceleration. To first approximation, any metric is flat. Therefore, to first approximation you are uniformly accelerating in a flat metric, which is precisely the physical meaning of the Rindler coordinates: $\rho = constant$ is the world line of a constantly accelerating object, and the $\tau = constant$ lines are the changing simultaneity surfaces of this accelerated observer.

• **The maximal extension**

Coordinates that cover the exterior, the black hole as well as the white hole, were found by Martin Kruskal and George Szekeres. They are (dimensionless) null coordinates U, V that go from $-\infty$ to $+\infty$, plus the usual angular coordinates θ, ϕ. Let's get to these coordinates in two steps. First, introduce the null coordinates

$$u = t - r - 2m \log |r/2m - 1|, \quad v = t + r + 2m \log |r/2m - 1|. \tag{11.19}$$

These cover only the exterior of the hole (they diverge on $r = 2m$) and put the Schwarzschild metric in a simple form:

$$ds^2 = -\left(1 - \frac{2m}{r}\right) du\, dv + r^2 d\Omega^2, \tag{11.20}$$

where $r = r(u, v)$ is understood as the function of u and v implicitly defined by

$$v - u = 2r + 4m \log |r/2m - 1|. \tag{11.21}$$

Now we need to extend the spacetime past $u = \infty$ and past $v = -\infty$. It suffices to define

$$U = -e^{-u/4m}, \quad V = e^{v/4m}. \tag{11.22}$$

In these coordinates, the metric reads

$$ds^2 = \frac{32m^3}{r} e^{-\frac{r}{2m}} \, dU \, dV + r^2 d\Omega^2, \tag{11.23}$$

where $r = r(U, V)$ is understood as the function of U and V defined (implicitly) by

$$UV = (2m - r) \, e^{\frac{r}{2m}}, \tag{11.24}$$

r is the Schwarzschild radius. The Schwarzschild time t is related to these coordinates by

$$t = 4m \, \text{arctanh} \, \frac{v - u}{v + u}. \tag{11.25}$$

To understand what is going on here, recall that the Minkowski metric can be put in the form (3.115), which is the analogue of (11.23). The change of coordinates (11.22) is the same we considered in (3.116) and the metric (3.117) is the analogue of (11.20). In both cases, the coordinates u, v cover only a limited portion of spacetime bounded by light surfaces, while the coordinates U, V cover the entire spacetime. To exit the Rindler wedge, we have to go 'across' $u = \infty$ (which is nothing else than $U = 0$ or $x = t$) or 'across' $v = \infty$ (which is nothing else than $V = 0$ or $x = -t$). Similarly, to enter the black hole or the white hole we have to go across the respective horizons, where the Schwarzschild coordinates diverge.

- **Carter–Penrose diagrams**

To understand the full extended geometry that these coordinates cover, let us draw the two-dimensional space formed by the radial and temporal direction. Every point represents thus a sphere of radius r. A convenient way to draw a Lorentzian space is to draw null coordinate lines always at ±45 degrees, so that the light cones are not deformed, and compress the infinite spacetime down to a finite region. This is always possible: the resulting picture reproduces distances very unfaithfully but reproduces causal relations faithfully. These spacetime diagrams are called 'Carter–Penrose diagrams', or 'Penrose diagrams', or 'conformal diagrams', because a squeezing that respects angles and the ±45-degree light cone structure is a conformal transformation. This can be easily done by defining the coordinates

$\tilde{U} \in [-\pi/2, \pi/2]$ and $\tilde{V} \in [-\pi/2, \pi/2]$ by

$$U = \tan \tilde{U}, \quad V = \tan \tilde{V} \qquad (11.26)$$

and plotting them as Cartesian coordinates of a two-dimensional Euclidean plane. Then the entire spacetime is mapped into a square. However, we have seen that the geometry becomes degenerate at $r = 0$; therefore, we need to restrict to the region of positive r. This gives the space of Figure 11.6, which shows the relative locations of the white hole region, the black hole region, and the exterior.

Remarkably, in this geometry there is a second exterior region, separated from the first. This second region has a separate asymptotic infinity.

Due to quantum phenomena, approaching $r = 0$ we leave the domain of validity of classical general relativity; therefore, we must limit ourselves to the regions of positive r. Note that the two regions $r = 0$ are spatial regions (in Minkowski $r = 0$ is a timeline). The

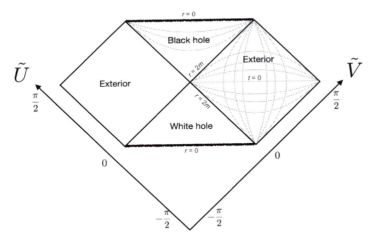

FIGURE 11.6 The maximal extension of the Schwarzschild geometry: the Kruskal space. In the exterior of the hole some lines of constant Schwarzschild radius and constant Schwarzschild time are depicted. Inside the black hole are shown some lines of constant Schwarzschild radius.

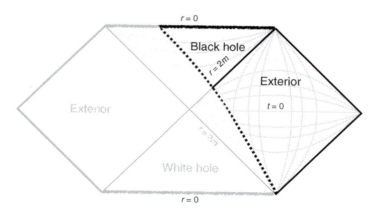

FIGURE 11.7 The part of the Kruskal spacetime that is relevant for the
geometry around a collapsing star. The dotted line is the surface of the
star.

geometry formed by the black and white holes, in addition to the
two outer regions, is the largest possible extension of Schwarzschild
spacetime as (pseudo) Riemannian geometry. Quantum theory could
extend it further, beyond $r = 0$.

• **Physical black holes formed by a collapsed star**
Most of the black holes we see in the sky were likely formed by
the collapse of a star, when the heat of nuclear fusion is no longer
sufficient to produce the pressure that counter-balances gravity.

Only the exterior of the star is described by a part of the
extended Schwarzschild spacetime described above, because the
Schwarzschild solution is a solution with $T_{ab} = 0$ and does not hold
inside the star, where the geometry is more simple.

Figure 11.7 illustrates the conformal diagram of a spacetime
with a collapsing star. In the beginning, there is only the star and
the outside. When the star enters the radius $r = 2m$, a horizon and a
trapped region are formed (i.e. the black hole).

Inside the black hole, the future reaches the $r = 0$ region. Again,
in the vicinity of $r = 0$ we leave the validity regime of classical theory.
The geometry of a collapsing star's black hole necessarily evolves in
a quantum region.

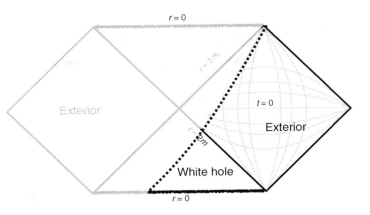

FIGURE 11.8 The part of the Kruskal spacetime that is relevant for the geometry around an exploding white hole. The dotted line is the surface of the exploding matter.

- **Exploding white holes**

The time reversal of a collapsing star is depicted in Figure 11.8. Notice that while the black hole *ends into* a quantum region, a white hole *emerges from* a quantum region.

While we have ample evidence that the geometry of Figure 11.7 describes actual phenomena in our universe, we have so far no direct evidence that the same holds for the geometry of Figure 11.8. So, physical white holes are hypothetical only for the moment. But so were black holes for quite a long time, after all.

The black hole geometry of a collapsing star ends into a quantum region. The white hole geometry emerges from a quantum region. It is possible to suspect that white holes could emerge from the same quantum region into which black holes end. I shall touch on this possibility in the last chapter, on quantum gravity.

12 Elements of Quantum Gravity

This final chapter provides a glimpse of the beautiful open problem of quantum gravity.

The domain of validity of classical general relativity is limited by the fact that the theory does not take quantum phenomena into account. The scale of the strength of quantum effects is governed by the Planck constant $\hbar$. Combining $\hbar$ with the Newton constant G and the speed of light c, we obtain constants with dimensions of Length, Time, Energy, Mass, and Density, called Planck length, Planck time, Planck energy, Planck mass, and Planck density:

$$L_{Pl} = \sqrt{\frac{\hbar G}{c^3}} \sim 10^{-33}\,\text{cm}, \qquad T_{Pl} = \sqrt{\frac{\hbar G}{c^5}} \sim 10^{-44}\,\text{s}, \qquad (12.1)$$

$$E_{Pl} = \sqrt{\frac{\hbar c^5}{G}} \sim 10^{19}\,\text{GeV}, \qquad M_{Pl} = \sqrt{\frac{\hbar c}{G}} \sim 20\,\mu\text{g}, \qquad (12.2)$$

$$\rho_{Pl} = \frac{c^5}{\hbar G^2} \sim 10^{93}\,\text{g/cm}^3. \qquad (12.3)$$

These constants set the scale for quantum gravity phenomena. They all play a role in this chapter.

Currently there is no quantum theory of gravity that has yet yielded genuinely new and empirically confirmed predictions. Therefore, the problem is open. One of the well-developed theoretical approaches focused on the problem of developing such a theory is *loop quantum gravity*, which will be mentioned below.

12.1 EMPIRICAL AND THEORETICAL BASIS OF QUANTUM GRAVITY

- *Empirical inputs*

The empirical basis of quantum gravity is solid: it is the huge body of facts that support classical general relativity and quantum mechanics. A theory of quantum gravity must be consistent with this vast domain of facts – namely with both quantum theory and general relativity where reliable – and must be internally consistent and sufficiently well defined to be predictive and testable. All this is far from easy: we do not bask in a wealth of possible alternatives. The tentative theories that are coming close to achieving this are rare.

In addition, we also have some *direct* empirical information that allows us to weight – and disfavour – some alternatives. Until a few years ago, the general impression was that quantum gravitational phenomena were too far outside our experimental reach. This feeling has receded recently, for a number of reasons:

- The energy reached by particle accelerators ($\sim 10^4\,\mathrm{GeV}$) is far from Planck energy ($E_{Pl} \sim 10^{19}\,\mathrm{GeV}$). But some particle physics experiments have tested phenomena at much higher energies, not far from E_{Pl}. For example, the appealing Georgi–Glashow $SU(5)$ theory has been disfavoured by showing that protons do not decay within the lifetime it predicts ($10^{30} - 10^{31}$ years). The super-Kamiokande experiment set a limit at $\sim 10^{33}$ years. Proton decay is a phenomenon at a scale $\sim 10^{16}\,\mathrm{GeV}$, only three orders of magnitude below E_{Pl}. Super-Kamiokande is not a collision apparatus, like particle accelerators: detectors surround a large number of protons – a tank of water. Smashing particles is not the only way to investigate high energies.

- Quantum gravity theories that break Lorentz invariance were vigorously explored a few years ago. This prompted an astronomical search for Lorentz-violating astrophysical effects, which has ended up largely excluding several such phenomena, at energies well above E_{Pl}, thus decreasing the plausibility of these tentative quantum gravity theories.

- Supersymmetric particles were expected to be observed at the LHC (Large Hadron Collider in Geneva) by some research communities, but they have been largely excluded at the LHC scale. This result does not falsify the quantum gravity research directions pursued by those communities, because supersymmetry could still be realised at higher energies, but it does reduce the expected probability that these programs will succeed, for the same (Bayesian) logic for which a positive result would have increased this probability.

- If successful, the gravity-induced entanglement experiment described in 12.3 below will rule out many ideas about geometry only being always, or macroscopically, classical.

- Loop quantum cosmology, mentioned below in 12.4, is exploring the possibility of computing effects that are possibly observable in the cosmic microwave background.

- Possible observational phenomenology of the black-to-white-hole transition, illustrated below in 12.4, has been explored. This includes dark matter, gamma rays, fast radio bursts, and others.

We do not yet have direct empirical evidence of a quantum gravitational phenomenon, but quantum gravity is definitely not disconnected from observations.

- *Theoretical basis*

We expect the gravitational field to exhibit quantum properties in the appropriate domain, as do all physical fields. But the quantum field theories effective for describing non-gravitational physics rely on the existence of a fixed metric structure of spacetime. This enters in virtually all the equations of a conventional quantum field theory. Once we take into account the relativistic properties of gravity, this structure itself becomes a quantum dynamical field. Hence the fixed metric used in the conventional construction of quantum field theory is not anymore available. It follows that most conventional quantum field methods are not appropriate for gravity. This difficulty is called the problem of the 'background independence' of quantum gravity.

In simpler terms, general relativity is not a field theory *on* a spacetime geometry, but rather the theory of the geometry *of* spacetime itself, and quantum gravity is not a quantum field theory *on* a spacetime geometry, but rather the quantum theory *of* the geometry of spacetime itself. Let see what this implies.

The distinctive features of any quantum theory are three:

1 *Discreteness*. Many physical variables take only discrete values in interactions. The electromagnetic field, for instance, interacts via discrete photons.
2 *Quantum superposition*. Generic quantum states do not correspond to classical configurations, but include 'linear superpositions' of these. These manifest themselves in the phenomenon of quantum interference.
3 *Probability*. The theory does not give sharp predictions of events, but only probability amplitudes for these.

Below I give some simple illustrative examples where these features appear in considering the quantum properties of geometry.

12.2 DISCRETENESS: QUANTA OF SPACE

In electromagnetism, photons are a characteristic manifestation of quantum discreteness: in a sense, photons show that the electromagnetic field is 'granular' at a small scale. Similarly, the gravitational field is granular at a small scale, and since the gravitational field *is* spacetime, this means that spacetime is granular at a small scale. The quantity that characterises quantum gravitational discreteness of space is the Planck length L_{Pl}.

To illustrate this granularity, consider a region R of a 3d Euclidean physical space whose geometry is described by a constant flat gravitational triad field e_a^i. For concreteness, imagine a region with the geometry of a tetrahedron with four vertices that we label as $A = 1, 2, 3, 4$. (Other geometries yield the same results.)

- *Classical geometry of a tetrahedron*

The geometry (including the size) of the tetrahedron is determined by the length of its six sides, which in turn are functions of the gravitational field. A convenient parametrisation of the geometry, which avoids the inequalities that these lengths have to satisfy, is given by the quantities

$$E_A^i = \frac{1}{2}\epsilon^i{}_{jk}\int_{\tau_A} e^j \wedge e^k, \tag{12.4}$$

where τ_A is the triangle opposite to the vertex A. The quantities E_A^i describe the geometry. They are four vectors normal to the faces whose length is equal to the area of the respective face. Indeed, it is easy to see *[show it!]* that the area of the triangle τ_A is given by

$$A_A = |E_A| \equiv \sqrt{\delta_{ij}E_A^i E_A^j}. \tag{12.5}$$

Compare with (3.107) and (3.108): E_A^i is the flux of the gravitational electric field across the face.

The geometry of the tetrahedron does not change for a rotation of the triad, which gives $E_A^i \mapsto R_j^i E_A^j$, where R is a rotation matrix. And it is easy to see (by Gauss' theorem) that *[show it!]*

$$\sum_A E_A^i = 0. \tag{12.6}$$

The volume of the tetrahedron is *[show it!]*

$$V = \frac{\sqrt{2}}{3}\sqrt{\epsilon_{ijk}E_A^i E_B^j E_C^k} \tag{12.7}$$

where A, B, and C are any three vertices of the tetrahedron ordered clockwise. Equation (12.6) implies that the choice of the vertices is irrelevant *[show it!]*.

- *Quantum geometry*

In quantum theory, the tetrad field is a quantum operator, and therefore the quantities E_a^i are quantum operators. Since they are vectors transforming in the vector representation of $SU(2)$, we may expect that they act on a Hilbert space that carries a representation of $SU(2)$. This is exactly what happens in loop quantum gravity, where the operators E_a^i are generators of $SU(2)$ that satisfy the $SU(2)$ commutation relations

$$[E_A^i, E_B^j] = c\delta_{AB}\,\epsilon_{ijk}\,E_a^k, \tag{12.8}$$

c is a constant of the dimension of a length squared, hence proportional to L_{Pl}^2. It is conventional to write this as

$$c = 8\pi\gamma L_{Pl}^2. \tag{12.9}$$

It can be shown that the commutation relations (12.8) reproduce ($\hbar$ times) the Poisson brackets of the classical theory, which can be derived via detailed canonical analysis from the action (5.3). (I do not show this here.) This derivation allows us to identify γ with the Barbero–Immirzi parameter introduced in (5.3). Therefore, these commutation relations are a conventional quantisation ansatz, like

$$[q, p] = i\hbar, \tag{12.10}$$

which reproduce ($\hbar$ times) the classical Poisson brackets and are the basic quantisation ansatz of quantum mechanics, à la Dirac.

The commutation relations (12.8) define a quantum theory of the geometry. Let's see what they imply.

• **The spectral analysis of the area and volume operators**
Equation (12.8) shows that for each A, the operators E_A^i are proportional to $SU(2)$ generators J^i, the well-known angular momentum operators

$$E_A^i = 8\pi\gamma L_{Pl}^2 J^i. \tag{12.11}$$

Therefore, the area operator is given by

$$A_A^2 = (8\pi\gamma L_{Pl}^2)^2\, J^i J^i = (8\pi\gamma L_{Pl}^2)^2\, L^2, \tag{12.12}$$

where L^2 is the Casimir of $SU(2)$, namely the well-known total angular momentum operator. This operator has discrete spectrum, and its eigenvalues are $j(j + 1)$ with $j = 0, 1/2, 1, 3/2, 2, \ldots.$ Therefore, we obtain the important result that the area operator has eigenvalues

$$A = \frac{8\pi\hbar G\gamma}{c^3} \sqrt{j(j + 1)} = 8\pi\gamma\, L_{Pl}^2 \sqrt{j(j + 1)}. \tag{12.13}$$

As this formula indicates, the Planck length gives the scale for the minimal size of quanta of physical space.

The eigenspaces that diagonalise the area operators are formed by four representations V_{j_A} of $SU(2)$, one per triangle A. A simple computation shows that the volume operator commutes with the area operators [do it!]. The volume operator therefore acts on the finite dimensional space

$$H = \otimes_A V_{j_A} \tag{12.14}$$

and can be diagonalised in each of these finite dimensional spaces, where it has a discrete spectrum as well, since these are finite dimensional Hilbert spaces.

The discrete spectra of the area and the volume operators capture the discreteness of space. They mean that physical space cannot be divided ad infinitum. The minimum measurable amounts of area and volume are finite and are at the Planck scale. The Planck length therefore expresses the finite divisibility of physical space.

- **Quanta of space**

An elementary 'quantum of space' is characterised by the discrete quantum numbers of the areas of its faces and of its volume. Notice that for a tetrahedron these are only five quantities, while the classical geometry of the tetrahedron is defined by six quantities. There is no sixth independent geometrical observable that commutes with area and volume; therefore, the six quantities that characterise the classical geometry cannot all be sharp.

The situation is the same as for angular momentum in elementary quantum mechanics: classical angular momentum is given by three quantities (L^x, L^y, L^z), but only two can be diagonalised together (L^z, L^2). Therefore, at the scale $\sim \hbar$ the angular momentum

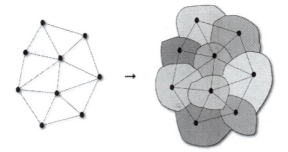

FIGURE 12.1 The graph of a spin network and the quanta of space it represents

components are discrete but never all sharp. In the same way, the geometry at the Planck scale is discrete but never entirely sharp.

- **Spin networks**

In loop quantum gravity, quantum states of spacetime are described by elements of a Hilbert space where a basis is given by networks of the quanta of space described above. The quanta of space form the nodes n of the network, and the links l connect adjacent nodes that share a face. Nodes and links form a 'graph' Γ, and the quantum states of this basis are written as

$$|\psi\rangle = |\Gamma, j_l, v_n\rangle, \qquad (12.15)$$

where j_l is the spin of the area of each face l, while v_n is the quantum number of the volume of each node n, and Γ codes the adjacency relations. The intuitive relation between the graph and the quanta of space is represented in Figure 12.1. These states are called *spin network states*.

These are basis states of quantum geometry, just as n-photon states are basis states in quantum electromagnetism.

The difference is in the fact that photon states live on a background spacetime, while spin network states define physical space themselves. This is reflected in the corresponding quantum numbers. The quantum numbers of a photon are the eigenvalues of its momentum, which is the Fourier transform of its position in a background physical space. The quantum numbers of a spin network state are the eigenvalues of the area and volume of the quanta of space of the state itself.[1]

12.3   SUPERPOSITION OF GEOMETRIES

Let us now consider a second aspect of quantum gravity: the superposition of geometries. A generic quantum state of spacetime is not a spin network state: it is a quantum superposition of these, as in any quantum theory. Spacetime geometries can be in quantum superposition.

I illustrate this phenomenon by discussing a proposed laboratory experiment.[2] The experiment aims at measuring a non-relativistic quantum gravitational effect, where spacetime geometries are put in quantum superposition, and quantum interference between these geometries is observed. This is called *gravitational entanglement*, or QGEM (Quantum Gravity induced Entanglement of Masses), or BMV effect (from Bose *et al.*, Marletto and Vedral). I describe it in a simple version.

- **The gravitational entanglement experiment**

The experiment is performed using two nanoparticles with spin and mass m. Both nanoparticles are split into a quantum superposition of two different positions (as in the classical Stern–Gerlach experiment); they are then both recombined after a time T. See Figure 12.2.

This generates four distinct branches of the quantum state for a time T. The idea of the experiment is to arrange positions so that in one of these four branches the two particles are kept at a small

[1]   See, for instance, C. Rovelli, 'Quantum gravity' (2004) and C. Rovelli, F. Vidotto, 'Covariant loop quantum gravity' (2014).

[2]   S. Bose *et al.* 'Spin entanglement witness for quantum gravity', *Phys. Rev. Lett.* 119, 240401 (2017). C. Marletto, V. Vedral, 'Gravitationally induced entanglement between two massive particles is sufficient evidence of quantum effects in gravity', *Phys. Rev. Lett.* 119, 240402 (2017).

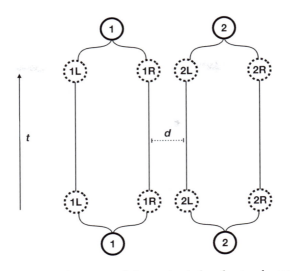

FIGURE 12.2 The setting of the gravity-induced entanglement experiment

distance d, so that each feels the gravitational field of the other. The effect of each particle is a slight deformation of the spacetime geometry in its vicinity. In each branch, the geometry can be approximated with the weak field formula (7.2) which here gives

$$ds^2 = -\left(1 + \frac{2\Phi_1(x)}{c^2} + \frac{2\Phi_2(x)}{c^2}\right)dt^2 + d\vec{x}^2, \qquad (12.16)$$

where the two terms are the Newtonian potential of the two particles. Consider the proper time along the trajectory of a particle. In all branches this is affected by the self-potential of the particle itself and by the potential due to the other particle. The self-potential is always the same, but the potential due to the other particle depends on the distance d between the particles

$$\frac{2\Phi_2(x)}{c^2} = -\frac{2Gm}{c^2d}, \qquad (12.17)$$

where d is the distance between the particles. The distance d is small in only one branch, in the other three we can disregard this term. The result is that, in the branch where the particles are close, the proper time

$$\int \sqrt{-ds^2} = \int_0^T dt \sqrt{1 - \frac{2Gm}{c^2 d}} \tag{12.18}$$

lapsed during the coordinate time T is shorter than in the other branches by the factor (compare with (7.10))

$$\delta T = \frac{Gm}{c^2 d} T. \tag{12.19}$$

The quantum state of a particle evolves with a phase $\exp\{imc^2 T/\hbar\}$. Therefore, when the branches recombine, the branch where the particles are kept nearby is out of phase with the others by a phase

$$\delta\phi = \frac{mc^2 \delta T}{\hbar} = \frac{Gm^2 T}{d\hbar}. \tag{12.20}$$

Imagine that the two particles were both in the plus eigenvalue of L_x and were split according to the eigenvalues of L_z. The split state is the tensor state

$$|\psi\rangle = (|+\rangle + |-\rangle) \otimes (|+\rangle + |-\rangle) = |++\rangle + |+-\rangle + |-+\rangle + |--\rangle. \tag{12.21}$$

If the branch where the two particles are nearby is the last one, then after a time T we have

$$|\psi(T)\rangle = |++\rangle + |+-\rangle + |-+\rangle + e^{i\delta\phi}|--\rangle. \tag{12.22}$$

If $\delta\phi = \pi$, this is a maximally entangled state (tracing $|\psi(\pi)\rangle\langle\psi(\pi)|$ on the first particle space state gives the identity on the second [show it!]), and this fact can be detected by measuring spins in repeated settings and checking the Bell inequalities. If the violation of these is verified, it follows that the two particles have been entangled by the effect of the gravitational potential.

It is not possible to entangle two quantum degrees of freedom if they only interact with a classical mediator. Hence the detection of the entanglement is a solid indication that the gravitational field is quantum: it was in a quantum superposition when the particles were so.

The key point of the experiment is that it reveals the superposition of the geometry. In fact, the entanglement happens only because the spacetime geometry is different in different branches: in each branch, it is given by (12.16), but d is different in each branch. Therefore, if this effect is verified, it may be taken as evidence that geometry can be in superposition.

• **Non-relativistic analysis**

The gravitational entanglement effect is non-relativistic, as is clear from the cancellation of c. It remains true in the limit in which $c \to \infty$. In fact, it can also be obtained from non-relativistic quantum mechanics by considering the gravitational pull as an instantaneous action at a distance. The branch where the particles are nearby has an additional Newtonian potential energy

$$\delta E = -\frac{Gm^2}{d}, \tag{12.23}$$

and the non-relativistic state evolves with the phase $\phi = e^{-iET/\hbar}$. This gives (12.20) again. Of course, in nature the interaction is not instantaneous and is mediated by the gravitational field: this knowledge is needed to draw the conclusion above.

This experiment, therefore, does not test the *relativistic* quantum gravity regime. A detection of the gravity-induced entanglement implies that spacetime is in a superposition of quantum geometries only in conjunction with the fact (which we know from general relativity) that the Newtonian gravitational potential is a manifestation of dynamical spacetime geometry.

• **Planck mass**

Notice that (12.20) can be written in the form

$$\delta\phi = \frac{m^2}{m_{Pl}^2} \frac{cT}{d}. \tag{12.24}$$

The second fraction is a dimensionless factor that characterises the setting of the experiment. The first fraction shows that the effect is governed by the Planck mass m_{Pl}. It is this mass that sets the scale at which particles of mass m can generate spacetime superpositions that can be observed in interference experiments like this one.

Today's technology allows us to bring nanoparticles of mass $m \sim 10^{-11}$ g in quantum superposition. It might be possible to hold them at a distance $d \sim 10^{-4}$ cm. Inserting this in (12.20) indicates that to have $\delta\phi \sim \pi$ we need $T \sim 1$ s. This might not be completely out of reach in the lab. If so, it might be possible to detect an effect of the quantum superposition of geometries in the lab in the coming years.

12.4   TRANSITIONS: BLACK-TO-WHITE-HOLE TUNNELLING AND BIG BOUNCE

The quantum evolution of the geometry must be described by transition amplitudes. In loop quantum gravity, the transition amplitude between spin network states is given by an expression based on the unitary representations of $SL(2, C)$. I shall not go into this here. Rather, I illustrate an application of the idea that the evolution of geometry can be given by quantum probability amplitudes.

• *Far future of black holes*

In Chapter 11, we left two problems open: what happens at the late stages of a black hole, when the geometry enters a quantum regime; and what happens in the early stage of a white hole, when the geometry emerges from a quantum region? Classical general relativity cannot answer these questions because curvature becomes Planckian at these stages, and the classical theory is not anymore reliable.

At the end of Section 11.2 we saw that this happens when the curvature reaches the Planck value, and this happens when

$$r \sim \sqrt[3]{\frac{G^2 \hbar M}{c^5}} = \sqrt[3]{\frac{M}{M_{Pl}}} \, L_{Pl}. \qquad (12.25)$$

Similarly, a collapsing star enters a quantum regime when the density is of the order of the Planck density, which occurs at this same

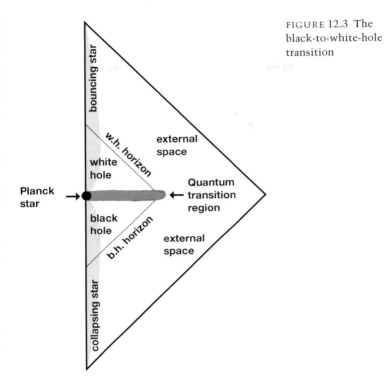

FIGURE 12.3 The black-to-white-hole transition

radius, because curvature and density are of the same order of magnitude (in natural units). For a macroscopic black hole $M \gg M_{Pl}$ and therefore $r \gg L_{Pl}$ (i.e. star and black hole enter a quantum regime with radii much larger than the Planck length L_{Pl}). A star that reaches this density is called a 'Planck star'.

What happens next?

A possibility that has been recently studied is that the late stage of a black hole can quantum tunnel into the early stage of a white hole.

The full spacetime including the transition can be represented by the Carter–Penrose diagram in Figure 12.3. The white region of the diagram satisfies the classical vacuum Einstein equations. (At first glance this is surprising, because this region cannot be *globally* immersed in the extended Schwarzschild metric: but it can be embedded *locally*.[3]) The dark grey region is the region where curvature is Planckian and therefore quantum gravity is relevant. In this region, the classical Einstein equations are violated, and we need to use quantum transition amplitudes to compute the probability for the transition to happen, as a function of the (parameters characterising the) geometry of the boundary of the transition region.

Calculations using the transition amplitudes of loop quantum gravity do indeed indicate that the transition probability becomes significative towards the end of the Hawking evaporation of the black hole. Hence it is possible that black holes tunnel into white holes towards the end of their evaporation.[4]

- **Quantum cosmology**

As we saw in Chapter 9, another physical situation where we exit the domain of validity of the classical theory is very early cosmology. Application of loop quantum gravity to the cosmological dynamics is called 'loop quantum cosmology'. A study of the approximation of the quantum dynamics (which I do not review here) leads to an effective Friedmann equation, which is modified at high density. Instead of (9.8), one obtains

$$\frac{\dot{a}^2}{a^2} = \frac{8}{3}\pi G \left(1 - \frac{\rho}{\rho_{max}}\right)\rho,$$ (12.26)

where

$$\rho_{max} = \frac{32\pi^2\gamma^3}{\sqrt{3}}\rho_{Pl}$$ (12.27)

is a constant of the order of the Planck density ρ_{Pl}.[5] The cosmological and the k terms of the Friedmann equation are negligible in the early universe.

---

[3] H. Haggard, C. Rovelli, 'Black hole fireworks: Quantum-gravity effects outside the horizon spark black to white hole tunneling', *Phys. Rev. D* (2015) 92.104020. arXiv:1407.0989.

[4] See, for instance, E. Bianchi *et al.*, 'White holes as remnants: A surprising scenario for the end of a black hole', *Class. Quant. Grav.* 35 (2018) no. 22, 225003. arXiv: 1802.04264, and references therein.

[5] See, for instance, I. Agullo, P. Singh, 'Loop quantum cosmology: A brief review', in *100 Years of General Relativity* (World Scientific 2017), arXiv:1612.01236.

The second term in the parentheses is the quantum correction. It becomes important only when the density is extremely high, and this happens in the very early universe, when the smallness of the scale factor compresses the matter energy. When this is so, this term counteracts the gravitational attraction and acts as a repulsive term, allowing the scale factor to bounce. The implication is that the 'Big Bang' might have followed a previously contracting phase of the universe and a 'Big Bounce' at the scale when the density has become Planckian. This may have left measurable traces in the cosmic microwave background.

12.5 CONCLUSION: THE DISAPPEARANCE OF SPACETIME

The few examples above give only a small (and partial) taste of the vast current research on quantum gravity. I hope they can give at least a hint of the beauty of this still very open field of research.

When combined with the discovery of the quantum nature of all physical variables, Einstein's everlasting insight that spacetime geometry is a manifestation of a dynamical field displays its full radicality.

Remember the distinction between the traditional *relational* notions of space and time, discussed in Section 2.1 and the new notions of space and time as independent entities that Newton introduced. The Newtonian notions of space and time, recognised as manifestations of the gravitational field by Einstein, further evolve into quantum variables in quantum gravity, showing that not just the fixed Euclidean geometry but the very notion of a continuous background geometry itself is only an approximation.

In this sense, space and time disappear from the conceptual structure of physics, following the same fate as the 'orbit of the electron' in the message Heisenberg sent to his friend Pauli after he had returned from Helgoland in 1925, with the key of quantum theory in his pocket: 'Everything is still very vague and unclear to me, but it seems that electrons no longer move in orbits.' On quantum gravity, 'Much is still vague and unclear to us, but it seems that spacetime is no longer there.'

But the gravitational field is still there, as a quantum field, with its quanta and computable transition amplitudes between its states.

Also, crucially, the old *relational* notions of space and time maintain their sense entirely. We can say that two events are adjacent, and we can count sequences of similar happenings: a clock is an object that does so. But there is no preferred clock in the theory nor any sense of localisation besides the relative adjacency of the quanta of matter and the quanta of space.

The shining beauty of Einstein's insight is still a source of knowledge and wonder.

Further Reading

This is a selection of entry points to the vast literature on general relativity.

The best discussion on the conceptual structure of general relativity is the Appendix 5 of Einstein's popular text on relativity:

- A. Einstein, *Relativity: The Special and the General Theory.* http://www.relativitycalculator.com/pdfs/relativity_the_special _ general_theory.pdf (© Eric Baird 1995 & 2008). (Unfortunately Appendix V is missing in several editions due to copyright issues.)

Introductions to general relativity

- R. D'Inverno, *Introducing Einstein's Relativity* (Oxford University Press, 1992)
- B. F. Schutz, *A First Course in General Relativity* (Cambridge University Press, 1985, 2009)
- J. B. Hartle, *Gravity, An Introduction to Einstein's General Relativity* (Addison Wesley, 2002; reissued by Cambridge University Press, 2021)
- L. Ryder. *Introduction to General Relativity* (Cambridge University Press, 2009)
- A. Barrau, *Relativité générale* (Dunot, 2011).

Classic manuals on general relativity

- S. M. Carroll, *Spacetime and Geometry* (Addison Wesley, 2004, reissued by Cambridge University Press, 2019)
- L. D. Landau, E. M. Lifshitz, *The Classical Theory of Fields* (Pergamon Press, 1971)

- R. M. Wald, *General Relativity* (University of Chicago Press, 1984)
- S. Weinberg, *Gravitation and Cosmology* (Wiley, 1972).

General relativity exercises

- T. A. Moore, *A General Relativity Workbook* (University Science Books, 2012).

Special relativity

- E. F. Taylor, J. A. Wheeler, *Spacetime Physics* (Freeman, 1992)
- A. M. Steane, *The Wonderful World of Relativity* (Oxford University Press, 2011).

Mathematics

- Y. Choquet-Bruhat, *Introduction to General Relativity, Black Holes and Cosmology* (Oxford University Press, 2015).
- T. Frankel, *The Geometry of Physics: An Introduction* (Cambridge University Press, 1997, 2004, 2011)

Quantum and gravity

- R. M. Wald, *Quantum Field Theory on Curved Spacetime* (Cambridge University Press, 1994)
- C. Rovelli, *Quantum Gravity* (Cambridge University Press, 2004)
- T. Thiemann, *Modern Canonical Quantum General Relativity* (Cambridge University Press, 2008)
- R. Gambini, J. Pullin, *A First Course in Loop Quantum Gravity* (Oxford University Press, 2013)
- C. Rovelli, F. Vidotto, *Covariant Loop Quantum Gravity* (Cambridge University Press, 2014).

Historical works cited in the text

- I. Newton, *Philosophiæ Naturalis Principia Mathematica* (Royal Society, 1687)
- C. G. Gauss 'Disquisitiones generales circa superficies curvas', auctore Carolo Friderico Gauss, *Societati regiae oblate* D.8. Octob 1827.
- B. Riemann, 'Über die Hypothesen, welche der Geometrie zu Grunde liegen', 'On the hypothesis on which geometry is based', *Abhandlungen der Königlichen Gesellschaft der Wissenschaften zu Göttingen*, vol. 13, 1867.
- A. Einstein, 'Die Feldgleichungen der Gravitation', *Sitzungsberichte der Preussischen Akademie der Wissenschaften zu Berlin*: 844–7, 1915.
- A. Einstein, 'Kosmologische Betrachtungen zur allgemeinen Relativitätstheorie', *Sitzungsberichte der Preussischen Akademie der Wissenschaften*: 142, 1917.
- A. Friedmann, 'Über die Krümmung des Raumes.' *Zeitschrift für Physik* 10(1):377–86, 1922. English translation in: Friedmann, A. (1999). 'On the curvature of space'. *General Relativity and Gravitation* 31 (12): 1991–2000.
- W. de Sitter, 'On the relativity of inertia: Remarks concerning Einstein's latest hypothesis', *Proc. Kon. Ned. Acad. Wet.* 19: 1217–25, 1917.
- R. Penrose, 'Gravitational collapse and space-time singularities', *Phys. Rev. Lett.* 14, 3 (1965).

Technical references in the last chapter

- S. Bose et al., 'Spin entanglement witness for quantum gravity', *Phys. Rev. Lett.* 119, 240401 (2017).
- C. Marletto, V. Vedral, 'Gravitationally induced entanglement between two massive particles is sufficient evidence of quantum effects in gravity', *Phys. Rev. Lett.* 119, 240402 (2017).

- H. Haggard, C. Rovelli, 'Black hole fireworks: Quantum-gravity effects outside the horizon spark black to white hole tunneling', *Phys. Rev. D* (2015) 92.104020. arXiv:1407.0989.
- E. Bianchi et al., 'White holes as remnants: A surprising scenario for the end of a black hole', *Class. Quant. Grav.* 35 (2018) no. 22, 225003. arXiv: 1802.04264, and references therein.
- I. Agullo, P. Singh, 'Loop quantum cosmology: A brief review', in *100 Years of General Relativity* (World Scientific, 2017), arXiv:1612.01236.

Popular introductory work to the general ideas of theoretical physics, general relativity and quantum gravity

- C. Rovelli, *Reality Is Not What It Seems* (Penguin, 2016)
- R. Gambini, J. Pullin, *Loop Quantum Gravity for Everyone* (World Scientific, 2020)
- J. Baggott, *Quantum Space: Loop Quantum Gravity and the Search for the Structure of Space, Time, and the Universe* (Oxford University Press 2020).

Index

Printed in the United States
by Baker & Taylor Publisher Services